Macmillan McGraw-Hill

Math Connects

5

Homework Practice
Workbook

Macmillan/McGraw-Hill

TO THE STUDENT This *Homework Practice Workbook* gives you additional examples and problems for the concept exercises in each lesson. The exercises are designed to help you study mathematics by reinforcing important skills needed to succeed in the everyday world. The materials are organized by chapter and lesson, with one Homework Practice worksheet for every lesson in *Math Connects, Grade 5*.

Always keep your workbook handy. Along with you textbook, daily homework, and class notes, the completed *Homework Practice Workbook* can help you in reviewing for quizzes and tests.

TO THE TEACHER These worksheets are the same ones found in the Chapter Resource Masters for *Math Connects, Grade 5*. The answers to these worksheets are available at the end of each Chapter Resource Masters booklet.

*The **McGraw·Hill** Companies*

 Macmillan/McGraw-Hill

Send all inquiries to:
Macmillan/McGraw-Hill
8787 Orion Place
Columbus, OH 43240

ISBN: 978-0-02-107299-6
MHID: 0-02-107299-X   *Homework Practice Workbook, Grade 5*

Printed in the United States of America.

2 3 4 5 6 7 8 9 10 047 14 13 12 11 10 09 08

CONTENTS

Name _____ Date _____

Homework Practice

Place Value Through Billions

Name the place value and write the value of the underlined digit.

1. 246,007,112

_____

_____

2. 712,409,625

_____

_____

3. 856,415

_____

_____

4. 19,003,017

_____

_____

Write each number in expanded form.

5. 23,618

_____

_____

6. 105,770,000

_____

_____

_____

7. 1,413,001,000

_____

_____

_____

8. 25,501,261

_____

_____

_____

1-2

Homework Practice

Compare Whole Numbers

Replace each ◯ with <, >, or = to make a true sentence.

1. 2,040 ◯ 20

2. 13,052 ◯ 16,912

3. 201 ◯ 2,001

4. 433,778 ◯ 433,778

5. 6,321 ◯ 282

6. 9 ◯ 13

7. 31 ◯ 38

8. 912 ◯ 921

9. 334 ◯ 3,340

10. 657 ◯ 567

Write the numbers in expanded form. Replace each ◯ with <, >, or = to make a true sentence.

11. 3,412 = _____

3,421 = _____

3,412 ◯ 3,421

12. 932 = _____

9,322 = _____

932 ◯ 9,322

Spiral Review

Write each number in standard form. (Lesson 1–1)

13. 11 million, 106 thousand, 300

_____

14. 91 billion, 13 million, 70 thousand, 2

_____

15. 300,000 + 20,000 + 300 + 70 + 8

_____

16. 1,000,000 + 50,000 + 9,000 + 4

_____

Name _____ Date _____

Homework Practice

Problem-Solving Investigation: Use the Four-Step Plan

Use the four-step plan to solve each problem.

1. A train left the station at 12:45. It traveled 455 miles in 7 hours. How many miles did it travel in each hour?

   _____

2. The Delgados are buying a pool that is 30 feet x 30 feet for $1,188. They plan to pay in 12 equal payments. Find the amount of each payment.

   _____

3. After shopping for school supplies, Martin came home with $4. He bought a pack of pens for $6, a calculator for $12, and a notebook for $3. How much money did he start with?

   _____

4. Julio increases the laps he runs by three laps each day. If he begins on Monday by running 4 laps, how many laps will he run on Wednesday at his current rate?

   _____

Spiral Review

Replace each ◯ with <, >, or = to make a true sentence. (Lesson 1–2)

5. 17 ◯ 8

6. 68 ◯ 93

7. 121 ◯ 1,210

8. 3,410 ◯ 3,401

9. 17,681 ◯ 16,681

10. 3,725,720 ◯ 3,752,720

Name _____ Date _____

Homework Practice

Represent Decimals

Write each fraction as a decimal.

1. $\frac{3}{5}$ _____

2. $1\frac{1}{4}$ _____

3. $\frac{37}{50}$ _____

4. $\frac{29}{100}$ _____

5. $\frac{127}{200}$ _____

6. $\frac{4}{5}$ _____

7. $6\frac{4}{25}$ _____

8. $\frac{19}{20}$ _____

9. $9\frac{1}{2}$ _____

10. $8\frac{7}{10}$ _____

11. $2\frac{7}{20}$ _____

12. $\frac{477}{500}$ _____

13. $\frac{129}{200}$ _____

14. $\frac{391}{500}$ _____

15. $\frac{493}{1,000}$ _____

Spiral Review

Solve. Use the four-step plan. (Lesson 1–3)

16. There are 15 students going to the museum. If each student pays $7 for admission and $5 for lunch, what is the total cost for the 15 students?

_____

17. Meredith worked on her sewing project for 45 minutes every night for 4 nights. She worked on the project for 30 minutes a night for the rest of the week. How many minutes did she work on her project altogether?

_____

Name _____ Date _____

Homework Practice

Place Value Through Thousandths

Write the place value and the value of each underlined digit.

1. 2.65<u>4</u> _____

2. 1.8<u>2</u> _____

3. 3.8<u>7</u> _____

4. 4.<u>9</u>5 _____

5. 12.93<u>1</u> _____

6. 135.<u>4</u>82 _____

Write each number in standard form.

7. 17 and 134 thousandths _____

8. 263 and 4 hundredths _____

9. 10 + 0.04 + 0.002 _____

10. 4+ 0.9 + 0.01 + 0.006 _____

Write each number in expanded form.

11. 174.273 _____

12. 30.024 _____

13. 209.106 _____

14. 44.815 _____

Spiral Review

Write each fraction as a decimal. (Lesson 1–4)

15. $\frac{3}{10}$ _____

16. $\frac{1}{10}$ _____

17. $\frac{67}{100}$ _____

18. $\frac{7}{100}$ _____

19. $\frac{413}{1,000}$ _____

20. $\frac{5}{1,000}$ _____

Homework Practice

Compare Decimals

Replace each ◯ with <, >, or = to make a true sentence.

1. 0.62 ◯ 0.618

2. 9.8 ◯ 9.80

3. 1.006 ◯ 1.02

4. 41.3 ◯ 41.03

5. 2.01 ◯ 2.011

6. 1.400 ◯ 1.40

7. 5.079 ◯ 5.08

8. 12.96 ◯ 12.967

9. 15.8 ◯ 15.800

10. 7.98 ◯ 7.89

11. 15 ◯ 15.01

12. 32.174 ◯ 32.740

13. 8.917 ◯ 8.907

14. 11.56 ◯ 11.5671

15. 0.01 ◯ 0.001

Spiral Review

Write each number in standard form. (Lesson 1–5)

16. 15 and 6 tenths _____

17. twenty and nine hundredths _____

18. 10 + 7 + 0.3 + 0.07 + 0.003 _____

19. 8 + 0.9 + 0.001 _____

Name _____ Date _____

Homework Practice

Order Whole Numbers and Decimals

Replace each ● with >, <, or = to compare each pair of decimals.

1. 0.788 ● 0.778 _____

2. 1.1 ● 1.10 _____

3. 4.052 ● 4.05 _____

4. 0.0549 ● 0.549 _____

5. 4.563 ● 0.4563 _____

6. 0.00783 ● 0.00837 _____

7. 9.34132 ● 9.31432 _____

8. 7.341 ● 70.041 _____

9. 0.30 ● 0.3000 _____

10. 1.8091 ● 1.8901 _____

11. 8.34 ● 8.43 _____

12. 0.23441 ● 0.34421 _____

13. 0.0120 ● 0.012 _____

14. 2.5038 ● 2.3058 _____

Order each set of decimals from *least* to *greatest*.

15. 2.654, 2.564, 2.0564, 2.465 _____

16. 1.11, 0.111, 1.01, 1.0011 _____

Spiral Review

Replace each ◯ with >, <, or = to make a true sentence. (Lesson 1-6)

17. 0.3 ◯ 0.2

18. 0.71 ◯ 0.17

19. 4.6 ◯ 4.60

20. 0.009 ◯ 1.09

21. 8.80 ◯ 8.88

22. 2.500 ◯ 2.5

Name _____ Date _____

Homework Practice

Problem-Solving Strategy: Guess and Check

Use the *guess and check* strategy to solve.

1. Jamal is thinking of four numbers from 1 through 9 whose sum is 21. Find the numbers.

   _____

2. Mr. Thompson took his 5 children to the amusement park. Tickets for children 12 and older cost $3.50. Tickets for children under 12 cost $2.25. He spends a total of $16.25. How many of his children are 12 and older?

   _____

3. A cabin has room for 7 campers and 2 counselors. How many cabins are needed for a total of 49 campers and 14 counselors?

   _____

Spiral Review

Order each set of numbers from *least* to *greatest*. (Lesson 1–7)

4. 147, 111, 175, 121 _____

5. 19.1, 15.3, 13.7, 18.5 _____

6. 0.83, 0.32, 0.88, 0.23 _____

7. 11,525; 11,125; 11,725; 11,225 _____

Name _____ Date _____

Homework Practice

Round Whole Numbers and Decimals

Round each decimal to the place indicated.

1. 1.583; ones _____

2. 67.095; tenths _____

3. 5.67; ones _____

4. 7.123; hundredths _____

5. 0.069; hundredths _____

6. 0.254; tenths _____

7. 569.8508; hundredths _____

8. 13.47; tens _____

9. 0.7010; thousandths _____

10. 10.89; tenths _____

11. 7.1385; thousandths _____

12. 0.571; hundredths _____

13. 215.073; hundreds _____

14. 105.148; tenths _____

Spiral Review

Solve. Use the *guess and check* strategy. (Lesson 1–8)

15. At the store, a pack of 10 baseball cards costs $1.25 and a pack of 15 baseball cards costs $1.75. José bought a total of 10 packs of baseball cards. If he spent a total of $14.50, how many of each size pack did he buy?

_____

_____

16. The Jackson family went to the local zoo. Children's tickets are $6.00 and adult tickets are $10.00. They spent $38.00 total on tickets. If they bought 5 total tickets, how many adult and children's tickets did they buy?

_____

Name _____ Date _____

Homework Practice

Estimate Sums and Differences

Estimate each sum or difference by rounding.

1. 5.30 + 1.76 + 4.079 _____

2. 2.341 − 1.51 _____

3. 100.38 − 16.653 _____

4. 2.462 + 6.90001 + 3.189 _____

5. 3.81 + 4.4913 _____

6. $6.46 + $5.50 _____

7. 1.62 + 2.40351 + 2.0099 _____

8. $4.82 + $5.51 + $5.01 _____

9. 60.032 + 5.2902 _____

10. $10.84 − $8.31 _____

11. $12.53 + $2.49 + $4.07 _____

12. 142.6447 − 44.9204 _____

Use the lunch menu for exercises 13–14.

13. Malcolm buys a taco and milk. About how much money does he spend?

_____

14. Susan buys a salad and two hamburgers. About how much does she spend?

_____

Lunch Menu	
Item	**Price**
Milk	$0.40
Hamburger	$1.25
Salad	$0.95
Taco	$1.49
Pudding	$0.70

Spiral Review

Round each decimal to the place indicated. (Lesson 2–1)

15. 6.1; ones _____

16. 9.23; tenths _____

17. 0.069; hundredths _____

18. 18.17; ones _____

19. 1.708; hundredths _____

20. 8.506; tenths _____

Name _____ Date _____

Homework Practice

Problem-Solving Strategy: Work Backward

Solve. Use the *work backward* strategy.

1. Ms. Houston's fifth grade class is going to a dinosaur park. The class raises $68 for the trip. Transportation to the park costs $40. The park sells small fossils for $4 each. How many fossils can they buy with the money they have left?

_____

2. The outdoors club went on a cross-country ski trip. Rentals for each person cost $4.50. Transportation for the group was $35. The total cost for rentals and transportation was $134. How many rentals did they pay for?

_____

3. Theresa had $15.65 left after a day at the mall. She spent $35 on a pair of running shoes, $12.50 on a shirt, and $3.85 on lunch. How much money did Theresa have when she arrived at the mall?

_____

4. Kusuo's baseball game begins at 5:00 P.M. Kusuo wants to arrive 45 minutes early to warm up. If it takes him $\frac{1}{2}$ hour to get to the baseball field, what time should Kusuo leave his home for the game?

_____

Spiral Review

Estimate each sum or difference by rounding. (Lesson 2–2)

5. $\begin{array}{r} 19 \\ + 12 \\ \hline \end{array}$

6. $\begin{array}{r} 398 \\ - 201 \\ \hline \end{array}$

7. $\begin{array}{r} 11.9 \\ + 6.8 \\ \hline \end{array}$

8. $1{,}397 + 115$

9. $1.69 - 0.71$

10. $42.38 - 21.28$

Name _____ Date _____

Homework Practice

Add and Subtract Whole Numbers

Add or subtract.

1.   58
      + 62

2.   301
      + 279

3.   913
      − 603

4.   708
      − 275

5.   4,025
      + 1,913

6.   8,642
      − 2,003

7.   21,745
      + 16,744

8.   489
      − 153

9.   608
      + 95

10.   6,245
       − 1,745

11.   968
       + 113

12.   12,729
       + 11,883

Spiral Review

Solve. Use the *work backward* strategy. (Lesson 2–3)

13. Russell wants to go bowling with his friend. Admission to the alley is $5 per person plus $2 for shoe rental and $3 an hour to bowl. If Russell has $20, and will also pay for his friend, how long can he bowl with his friend? _____

14. Jill finished her Saturday chores at 3 P.M. She spent 23 minutes washing the dog, 26 minutes vacuuming, 11 minutes folding laundry, and 30 minutes cleaning her room. What time did she start her chores? _____

2-5

Homework Practice

Problem-Solving Investigation: Estimate or Exact Answer

For each problem, determine whether you need an estimate or an exact answer. Then solve.

1. It costs Matt a little more that $4 a day to feed his dog. How much does it cost him to feed his dog for a year?

   _____

2. In the past year, a grocery store deposited about 6 million pennies, 3 million nickels, 4 million dimes, and 2 million quarters in the bank. What is the total value of the deposit?

   _____

3. A bank puts 3,000 quarters in each bag. How much are 15 bags of quarters worth?

   _____

4. A vault contains $3,000 worth of nickels. How many nickels are in the vault?

   _____

5. When at rest, your heart probably beats about 70 times per minute. At that rate, how many times does it beat in an hour?

   _____

6. Ann bought two shirts for $28.95 each and a skirt for $33.95. The sales tax was $3.71. How much did she pay altogether?

   _____

Spiral Review

Add or subtract. (Lesson 2-4)

7. $52 + 21$ _____

8. $1,026 + 962$ _____

9. $401 + 102$ _____

10. $17,342 - 9,603$ _____

11. $18,998 - 7,426$ _____

12. $65,017 - 8,406$ _____

Name _____ Date _____

Homework Practice

Add and Subtract Decimals

Add.

1. 1.546 + 0.07 _____

2. 75.4 + 0.5919 _____

3. $100.80 + $5.87 _____

4. 3.72 + 6.0064 _____

5. 0.802 + 0.4581 _____

6. 4.3 + 0.1748 _____

7. $0.35 + $0.54 _____

Subtract.

8. $11.11 – $4.88 _____

9. 5 – 4.825 _____

10. 10 – 3.485 _____

11. 16.2445 – 3.77 _____

12. 6.5 – 3.001 _____

13. 3.48 – 3.039 _____

14. 2.1 – 1.34 _____

Spiral Review

For each problem, determine whether you need an estimate or an exact answer. Then solve it. (Lesson 2–5)

15. Hunter School has kindergarten and grades 1 through 6. There are 2 kindergarten classes and 2 classes in each grade. If the maximum class size is 25, what is the greatest number of students that could be in the school?

_____

16. It costs $0.38 to produce and mail a newsletter. Each week, 475,000 newsletters are mailed to subscribers. About how much does it cost to produce and mail the newsletter each month?

_____

Name _____ Date _____

Homework Practice

Addition Properties

Use properties of addition to find each sum mentally. Show your steps and identify the properties that you used.

1. 9 + 31 + 6 _____

_____

_____

2. 17 + 23 + 8 _____

_____

_____

3. 12.5 + 1 + 43.5 _____

_____

_____

4. 10.4 + 20 + 9.6 _____

_____

_____

For Exercises 5–8, find the value that makes the sentence true.

5. 33 + (76 + 5) = ☐ + 33) + 5

6. 13.8 + (9.9 + 4) = (9.9 + 4) + ☐

7. 56 + (26 + 5) = (5 + ☐) + 56

8. 2.8 + (3.6 + 8.2) = (3.6 + 2.8) + ☐

Spiral Review

Estimate. Then add or subtract. (Lesson 2–6)

9. 13 + 9.25 = _____

10. 7.41 − 1.3 = _____

11. 18.21 − 10.03 = _____

12. 6.007 + 2.31 = _____

13. 9.8 − 0.19 = _____

14. 31.1 + 8.169 = _____

2-8

Homework Practice

Add and Subtract Mentally

Add or subtract mentally. Use compensation.

1. $18 + 26$ _____

2. $2.7 + 3.5$ _____

3. $134 - 78$ _____

4. $32.5 - 17.8$ _____

5. $221 + 335$ _____

6. $9.6 - 8.9$ _____

7. $129 + 375$ _____

8. $207 + 45.9$ _____

9. $754 - 543$ _____

10. $15.7 - 12.6$ _____

11. $553 + 35$ _____

12. $1.5 - 0.6$ _____

Spiral Review

For Exercises 13–17, find the value that makes the sentence true. (Lesson 2–7)

13. $50 + (67 + 8) = (\boxed{} + 67) + 50$

14. $12.4 + (9.7 + 3.4) = (3.4 + 9.7) + \boxed{}$

15. $19 + (21 + 3) = (21 + \boxed{}) + 3$

16. $5.7 + (2.3 + 3.2) = (2.3 + 5.7) + \boxed{}$

Name _____ Date _____

Homework Practice

Multiplication Patterns

Find each product mentally.

1. 6 × 100 = _____

2. 8 × 300 = _____

3. 20 × 50 = _____

4. 4 × 600 = _____

5. 1,000 × 23 = _____

6. 900 × 20 = _____

7. 800 × 60 = _____

8. 12 × 60 = _____

9. 12 × 5,000 = _____

10. 500 × 90 = _____

11. 11 × 300 = _____

12. 70 × 600 = _____

13. 60 × 50 = _____

14. 80 × 200 = _____

15. 90 × 70 = _____

16. 100 × 90 = _____

17. 600 × 12 = _____

18. 40 × 90 = _____

19. 50 × 700 = _____

20. 70 × 300 = _____

21. 40 × 80 = _____

22. 70 × 110 = _____

Spiral Review

Add or subtract mentally. Use compensation. (Lesson 2–8)

23. 24 + 56 = _____

24. 33 − 12 = _____

25. 49 + 62 = _____

26. 19 + 9 = _____

27. 57 − 38 = _____

28. 310 − 218 = _____

29. 589 + 221 = _____

30. 46 + 26 = _____

31. 39 + 61 = _____

32. 472 − 28 = _____

Name _____ Date _____

Homework Practice

The Distributive Property

Rewrite each expression using the Distributive Property. Then evaluate.

1. $3 \times (40 + 6)$

2. $6 \times (60 + 5)$

_____

_____

Find each product mentally using the Distributive Property. Show the steps that you used.

3. 28×6

4. 6×34

5. 35×7

6. 3×72

7. Mrs. Robertson bought 7 tickets for the school play on Monday and 5 tickets on Tuesday. Each ticket was $6. How much did she spend on the tickets? Show how you can use the Distributive Property.

8. In each package of school supplies there are 3 notebooks and 2 pencils. If you have 42 packages, how many notebooks and pencils do you have altogether? Use the Distributive Property and show your steps.

Spiral Review

Find each product mentally. (Lesson 3–1)

9. 10×18

10. 24×60

11. 300×9

12. 50×15

Name _____ Date _____

Homework Practice

Estimate Products

Estimate by rounding. Show your work.

1. 542
  × 38

2. 821
  × 14

3. 726
  × 38

4. 174
  × 73

5. 2,862
  × 143

6. 12,649
  ×   382

7. 69,238
  ×   54

8. 10,405
  ×   632

9. 14 × 77 _____

10. 34 × 873 _____

11. 469 × 18 _____

12. 89 × 806 _____

13. 47 × 962 _____

14. 3,721 × 499 _____

15. 198 × 2,203 _____

16. 3,926 × 198 _____

Spiral Review

Rewrite each expression using the Distributive Property. Then evaluate. (Lesson 3–2)

17. 6 × (10 + 5)

_____

18. 4 × (30 + 6)

_____

19. 3 × (30 + 8)

_____

20. 7 × (40 + 6)

_____

21. 2 × (10 + 7)

_____

22. 6 × (40 + 8)

_____

Name _____ Date _____

Homework Practice

Multiply by One-Digit Numbers

Multiply.

1. 47×3 **2.** 28×2 **3.** 65×5 **4.** 41×4

_____   _____   _____   _____

5. 6×37 **6.** 25×8 **7.** 94×7 **8.** 4×38

_____   _____   _____   _____

9. 249×6 **10.** 326×2 **11.** 3×547 **12.** 683×3

_____   _____   _____   _____

13. 552
    $\times \ \ 3$

14. 243
    $\times \ \ 4$

15. 671
    $\times \ \ 7$

16. 342
    $\times \ \ 2$

17. 128
    $\times \ \ 6$

18. 444
    $\times \ \ 5$

19. 831
    $\times \ \ 7$

20. 756
    $\times \ \ 2$

Spiral Review

Estimate by rounding. Show your work. (Lesson 3–3)

21. 107
    $\times \ 54$

22. 38
    $\times \ 7$

23. 602
    $\times \ 14$

24. 68
    $\times \ 69$

25. 42
    $\times \ 51$

26. 216
    $\times \ 7$

27. 19
    $\times \ 19$

28. 401
    $\times \ 33$

Name _____ Date _____

Homework Practice

Problem-Solving Strategy: Draw a picture

Solve. Use the *draw a picture* strategy.

1. Gregory arranges eight identical cubes into one large cube on a table. How many sides of the small cubes can he see?

_____

2. Al walks 3 blocks north, 2 blocks east, 3 blocks north, and 3 blocks east to get to the theater. Will the path home be any shorter if he walks south to his street, then west to his house? Why?

_____

3. Sandy's classroom has tables that are shaped like rectangles. Three of the tables are placed together in a U-shape. One student can sit at the short side of each rectangle or at the end of a table. Two students can sit at the long side of each rectangle. No students sit along the "inside" of the U-shape. How many students can sit at the tables the way they are arranged?

_____ students

4. Valerie is making coasters for her mother's craft booth. For each coaster, she uses a square of cork and four pieces of wood to glue along the edges. If she has 25 pieces of cork and 92 pieces of wood, how many coasters can she make?

_____

Will she run out of the cork first or the wood?

_____

Spiral Review

Estimate. Then Multiply. (Lesson 3–4)

5.   68
     × 4
     ____

6.   59
     × 6
     ____

7.   519
     ×   3
     ____

8.   874
     ×   2
     ____

9.   338
     ×   5
     ____

10.   902
      ×   3
      ____

Name _____ Date _____

Homework Practice

Multiply by Two-Digit Numbers

Multiply.

1. 142 × 65 = _____

2. 407 × 73 = _____

3. $396 × 84 = _____

4. 862 × 29 = _____

5. 64 × 981 = _____

6. 69 × 46 = _____

7. 57 × $37 = _____

8. 656 × 23 = _____

9. 390 × 48 = _____

10. 357 × 54 = _____

11. 378 × 76 = _____

12. 476 × 93 = _____

13. 73 × $547 = _____

14. 326 × 57 = _____

15. 318 × 21 = _____

16. 215 × 58 = _____

17. 19 × $739 = _____

18. 862 × 12 = _____

19. 84 × 119 = _____

20. 37 × 208 = _____

21. 239 × 17 = _____

22. 926 × 60 = _____

23. 85 × 63 = _____

24. 209 × 75 = _____

25. 45 × 306 = _____

26. 443 × 19 = _____

Spiral Review

Solve. Use the *draw a picture* strategy. (Lesson 3–5)

27. Tulips are planted every 4 feet around the outside edge of a rectangular garden. If the sides of the garden measure 16 feet and 12 feet, how many total tulips are there?

_____

28. Jill has 5 pictures to hang on her wall. She wants to hang one picture in the center and the other 4 at the corners of the center picture. If the picture in the center remains the same, how many different ways can she hang the other pictures?

Name _____ Date _____

Homework Practice

Multiplication Properties

Identify the multiplication property used to rewrite each problem.

1. $(81 \times 4) \times 5 = 81 \times (4 \times 5)$

_____

2. $(28 \times 7) \times 16 = 28 \times (7 \times 16)$

_____

3. $15 \times 8 = 8 \times 15$

_____

4. $72 \times 1 = 72$

_____

Use properties of multiplication to find each product mentally. Show your steps and identify the properties that you used.

5. $76 \times 25 \times 4$

_____

_____

_____

_____

6. $5 \times 60 \times 20$

_____

_____

_____

_____

7. $40 \times 0 \times 17$

_____

_____

_____

8. $2 \times 4 \times 12$

_____

_____

_____

Multiply. (Lesson 3–6)

9. 560
  $\times\ 24$

10. 68
  $\times\ 17$

11. 341
  $\times\ 25$

12. 18
  $\times\ 49$

13. 836
  $\times\ 32$

14. 95
  $\times\ 73$

15. 188
  $\times\ 58$

16. 712
  $\times\ 65$

Name _____ Date _____

Homework Practice

Extending Multiplication

Estimate each product.

1.	$5.27	2.	$3.36	3.	$12.17
	× 6		× 4		× 7

4.	$4.28	5.	$8.17	6.	$1.32
	× 5		× 8		× 5

7.	$27.64	8.	$63.44	9.	$17.55
	× 3		× 6		× 9

10.	7.7	11.	11.9	12.	51.7
	× 4		× 21		× 9

13.	33.3	14.	87.2	15.	17.6
	× 33		× 41		× 51

16.	27.1	17.	72.1	18.	92.1
	× 205		× 51		× 11

Spiral Review

Use properties of multiplication to find each product mentally. Show your steps and identify the properties that you used. (Lesson 3–7)

19. $5 \times 14 \times 2$

20. $50 \times 6 \times 20$

_____   _____

_____   _____

_____   _____

Name _____ Date _____

Homework Practice

Problem-Solving Investigation: Extra or Missing Information

Solve each problem. If there is extra information, identify it. If there is not enough information, tell what information is needed.

1. The Alvarez family bought a car for $2,000. They made a down payment of $500. If they want to pay the balance in five equal payments, how much will each of these payments be?

   _____

2. Melanie can walk five miles in a half hour. How many miles will she walk in one week?

   _____

3. Jamie and Melissa both have 30 CDs. How many more CDs does John have?

   _____

4. Anna collected 50 cans for a food drive. She collected 10 cans each day of the drive, mostly green beans. How many days did she collect cans?

   _____

Spiral Review

Estimate.

5. $6.82
    $\times \quad 4$

6. $3.37
    $\times \quad 2$

7. $41.64
    $\times \quad 12$

8. $35.61
    $\times \quad 24$

9. $17.52
    $\times \quad 18$

10. 2.67
    $\times \quad 54$

11. 94.2
    $\times \quad 76$

12. 76.4
    $\times \quad 127$

13. 8.6
    $\times \quad 82$

Name _____ Date _____

Homework Practice

Division Patterns

Divide mentally.

1. $270 \div 3 =$ _____
2. $480 \div 60 =$ _____
3. $180 \div 9 =$ _____

4. $2,000 \div 10 =$ _____
5. $300 \div 20 =$ _____
6. $400 \div 4 =$ _____

7. $560 \div 7 =$ _____
8. $3,200 \div 80 =$ _____
9. $600 \div 30 =$ _____

10. $4,500 \div 5 =$ _____
11. $8,100 \div 90 =$ _____
12. $600 \div 2 =$ _____

13. $2,800 \div 7 =$ _____
14. $1,800 \div 30 =$ _____
15. $500 \div 8 =$ _____

Solve.

16. Peyton has collected 120 aluminum cans for recycling. If 20 cans will fit in each blue plastic bag, how many bags will she need to carry all the cans?

_____

Spiral Review

Solve each problem. If there is extra information, identify it. If there is not enough information, tell what information is needed.

17. Kelly is making sandwiches for a picnic. She has ham, tuna, and cheese. How many loaves of bread will she need to make 4 sandwiches of each kind?

_____

18. Jake is building a birdhouse out of wood. Each side of the birdhouse will measure about one square foot. The roof panels will measure about 1.5 square feet total. He wants to attract robins and blue jays. How much wood will Jake need to build the birdhouse?

_____

Name _____ Date _____

Homework Practice

Estimate Quotients

Estimate. Show your work.

1. $231 \div 6$ _____

2. $149 \div 4$ _____

3. $4,748 \div 7$ _____

4. $275 \div 4$ _____

5. $314 \div 6$ _____

6. $5,603 \div 9$ _____

7. $8\overline{)629}$ _____

8. $9\overline{)290}$ _____

9. $9\overline{)342}$ _____

10. $5\overline{)9,461}$ _____

11. $8\overline{)2,943}$ _____

12. $7\overline{)33,875}$ _____

Solve.

13. Each of the 9 parking lots at an automobile plant holds the same number of new cars. The lots are full. If there are 4,131 cars in the lots, about how many cars are in each lot? Show your work.

_____

14. A total of 176 valves were used for 8 cars as they were being assembled. About how many valves were used for each car? Show your work.

_____

Spiral Review

Divide mentally. (Lesson 4–1)

15. $360 \div 6 =$ _____

16. $4,500 \div 5 =$ _____

17. $8,000 \div 100 =$ _____

18. $400 \div 8 =$ _____

19. $180 \div 30 =$ _____

20. $1,600 \div 400 =$ _____

21. $4,900 \div 7 =$ _____

22. $5,400 \div 60 =$ _____

23. $7,200 \div 80 =$ _____

Name _____ Date _____

Homework Practice

Divide by One-Digit Numbers

Divide.

1. $5\overline{)106}$

2. $7\overline{)862}$

3. $9\overline{)775}$

4. $3\overline{)195}$

5. $8\overline{)451}$

6. $6\overline{)918}$

7. $6\overline{)310}$

8. $3\overline{)803}$

9. $5\overline{)369}$

10. $4\overline{)701}$

11. $6\overline{)380}$

12. $6\overline{)102}$

Solve.

13. A family of 4 spent $64 for tickets to a soccer game. All of the tickets were the same price. What was the cost of each ticket?

_____

14. $350 was raised at a car wash. How many cars were washed if it costs $5 to wash one car?

_____

Spiral Review

Estimate. Show your work. (Lesson 4–2)

15. $4,875 \div 82 =$ _____

16. $2,602 \div 37 =$ _____

17. $2,148 \div 62 =$ _____

18. $8,932 \div 451 =$ _____

19. $3,494 \div 349 =$ _____

20. $9,456 \div 295 =$ _____

21. $27,568 \div 9 =$ _____

22. $5,688 \div 34 =$ _____

Name _____ Date _____

Homework Practice

Divide by Two-Digit Numbers

Divide.

1. $54\overline{)106}$

2. $17\overline{)862}$

3. $29\overline{)775}$

4. $13\overline{)195}$

5. $98\overline{)620}$

6. $26\overline{)918}$

7. $66\overline{)310}$

8. $53\overline{)803}$

9. $45\overline{)369}$

10. $14\overline{)701}$

11. $36\overline{)380}$

12. $54\overline{)710}$

Solve.

13. A ticket seller collected $990 for selling tickets. Each ticket costs $15. How many tickets did she sell?

_____

14. DVDs cost $20 each. How many DVDs can Miguel buy for $180?

_____

Spiral Review

Estimate. Then divide. (Lesson 4–3)

15. $975 \div 8 =$ _____

16. $702 \div 7 =$ _____

17. $248 \div 6 =$ _____

18. $832 \div 4 =$ _____

19. $594 \div 3 =$ _____

20. $556 \div 9 =$ _____

21. $668 \div 9 =$ _____

22. $788 \div 3 =$ _____

Name _____ Date _____

Homework Practice

Problem-Solving Strategy: Act It Out

Solve. Use the *act it out* strategy.

1. Alberto has 2 quarters, 5 dimes, and 18 nickels. How many different combinations of coins can he make to have $2?

   _____

2. Carlos is running drills of 0.5 mile. If he runs 5 drills, how many miles did he run?

   _____

3. Students are hanging their art projects in the school hallway. Each student wants to hang a project that is 12 inches wide. The hallway is 16 feet long. If they don't leave any spaces between projects, how many projects will fit in the hallway?

   _____

4. Hana is wrapping books to give as gifts. She needs pieces of wrapping paper that are 0.5 foot long for each book. She has a total of 6 books. How many feet of wrapping paper will she need?

   _____

Spiral Review

Divide. (Lesson 4–4)

5. $35\overline{)321}$ _____

6. $94\overline{)456}$ _____

7. $24\overline{)482}$ _____

8. $12\overline{)121}$ _____

9. $15\overline{)515}$ _____

10. $28\overline{)633}$ _____

11. $14\overline{)241}$ _____

12. $39\overline{)784}$ _____

13. $15\overline{)180}$ _____

Name _____ Date _____

Homework Practice

Interpret the Remainder

Solve. Explain how you interpreted the remainder.

1. There are 10 pencils to be given to 3 students. How many pencils does each student get?

   _____

   _____

   _____

2. Four employees of Papa Tony's Pizza are cleaning up at the end of a busy night. There is a list of 31 clean-up tasks that need to be completed. If each employee does the same number of tasks, how many tasks should each employee do?

   _____

   _____

3. Frank, Kendra, and Lin are painting a mural. There are 2 sets of paints, each with 50 colors. Paints are passed out until everyone has the same number. How many colors does each worker get?

   _____

   _____

4. Mr. Hollings gives small pieces of fruit to his class while students work on their history projects. He has 77 pieces of fruit, and 9 students in his class. How many pieces of fruit will each student get?

   _____

   _____

Spiral Review

Solve. Use the *act it out* strategy. (Lesson 4–5)

5. Ron has 5 new songs on his MP3 player. In how many different orders can he burn them to a CD?

   _____

6. Sixty-seven students are going on a field trip to the zoo. The principal needs to decide how many cars and vans are needed to transport the students. Each car can seat 5 students and each van seats 9. The principal would like to use the fewest number of vehicles possible. How many cars and how many vans should she use?

   _____

Name _____ Date _____

4-7

Homework Practice

Extending Division

Estimate each quotient.

1. 18.4 ÷ 6 = _____

2. 32.1 ÷ 6 = _____

3. 50.3 ÷ 7 = _____

4. 26.6 ÷ 9 = _____

5. 12.3 ÷ 5 = _____

6. 10.8 ÷ 5 = _____

7. 17.5 ÷ 5 = _____

8. 120.6 ÷ 2 = _____

9. 32.1 ÷ 8 = _____

10. 88.9 ÷ 3 = _____

11. 31.6 ÷ 5 = _____

12. 41.5 ÷ 5 = _____

13. $7.80 ÷ 6 = _____

14. $89.10 ÷ 4 = _____

15. $26.14 ÷ 4 = _____

16. $7.30 ÷ 5 = _____

17. $15.12 ÷ 8 = _____

18. $33.39 ÷ 6 = _____

Spiral Review

Solve. Explain how you interpreted the remainder. (Lesson 4–6)

19. A package of 25 pencils is divided among 10 students. How many pencils does each student get?

_____

20. One car can seat 5 people. If Kate's parents take her and 6 friends to the movies, how many cars will be needed?

_____

Name _____ Date _____

Homework Practice

Problem-Solving Investigation: Choose the Best Strategy

Solve. Use any strategy shown below to solve each problem.
- **Draw a picture**
- **Work backward**
- **Guess and check**
- **Act it out**

1. Lydia planted beans in her garden. After a week they grew 3 inches. The next week they measured 5 inches. In the third week they measured 7 inches. If they continue growing at this rate, how tall will the beans be in the fifth week?

_____

2. The restaurant is expecting a group of 84 people for dinner. Each table in the restaurant seats 6 people. If the restaurant has 15 tables, will they have enough tables for the large group? Explain your answer.

_____

3. The fifth grade art class is creating a mural for the library. The mural is 28 feet long and divided into 7 sections. How many feet long is each section of the mural?

_____

4. Olivia is setting up a window display for the grocery store. The first row has 8 pieces of fruit, the second row has 12 vegetables, and the third row has 16 loaves of bread. How many canned goods are in the fourth row?

_____

Spiral Review

Estimate each quotient. (Lesson 4–7)

5. $4.50 ÷ 5 = _____

6. $72.30 ÷ 9 = _____

7. 9.4 ÷ 5 = _____

8. 13.3 ÷ 2 = _____

9. 6.12 ÷ 2 = _____

10. $23.80 ÷ 7 = _____

11. 130.2 ÷ 6 = _____

12. $19.70 ÷ 2 = _____

13. $21.13 ÷ 5 = _____

14. 38.51 ÷ 5 = _____

Name _____ Date _____

Homework Practice

Addition Expressions

Evaluate each expression if x = 7 and y = 3.

1. $x + 5$ _____

2. $10 + y$ _____

3. $4 + x$ _____

4. $y + 12$ _____

5. $3 + x$ _____

6. $11 + y$ _____

7. $15 + x$ _____

8. $y + 19$ _____

9. $y + 8$ _____

Write an expression for each real-world situation. Then evaluate.

10. Annie had x dollars. Her dad gave her $25. If $x = 10$, how much money does Annie have?

11. Warren jogged 5 miles. His friend jogged y miles farther. If $y = 2$, how many miles did Warren's friend jog?

 **Spiral Review**

Use any strategy shown below to solve each problem. (Lesson 4–8)

- Act it out.
- Draw a picture
- Work backward
- Guess and check

12. Ken bought CDs and a pair of socks. He spent $3.95 on the socks and $12.99 for each CD. The items cost $42.92 before tax. How many CDs did Ken buy?

13. Paula needs to catch the school bus by 8:45 A.M. It takes her 20 minutes to shower, 10 minutes to dress, and 15 minutes to eat breakfast. What time does Paula need to wake up?

Homework Practice

Problem-Solving Strategy: Solve a Simpler Problem

Solve. Use the *solve a simpler problem* strategy.

1. Noelle has a deck that extends 4 ft on all sides of an 18 ft by 10 ft rectangular swimming pool. What is the area of the deck?

_____

2. Michaella plants flowers in a rectangular pattern. The pattern has a length of 6 ft and a width of 5 ft. If a tree takes up a square that is 1 ft on a side in the middle of the pattern, what area is planted in flowers?

_____

3. Six bakers can make six cakes in six hours. How many cakes can twelve bakers make in twelve hours?

_____

4. Miguel earns $50 each week delivering newspapers. He keeps $10 each week and puts the remaining amount into his savings account. How many weeks will it take until he has more than $300 in his savings account?

_____

5. Liem wants to hang a picture on his bedroom wall. The picture itself is 4 inches by 6 inches, and the frame extends 2 inches around the picture. How much wall space will the framed picture take?

_____

Spiral Review

Evaluate each expression if $x = 3$ and $y = 7$. (Lesson 5–1)

6. $x + 3$ _____

7. $7 + x$ _____

8. $1 + y$ _____

9. $y + 8$ _____

10. $10 + x$ _____

11. $12 + y$ _____

12. $2 + x$ _____

13. $y + 25$ _____

14. $x + 36$ _____

Name _____ Date _____

Homework Practice

Multiplication Expressions

Evaluate each expression if *x* = 3 and *y* = 5.

1. 5*x* _____

2. 9*y* _____

3. 2*x* _____

4. 11*y* _____

5. 7*x* _____

6. 4*y* _____

Evaluate each expression if *a* = 7 and *b* = 6.

7. 3*a* _____

8. 12*b* _____

9. 14*a* _____

10. 8*a* _____

11. 22*b* _____

12. 6*b* _____

Solve. Use the *solve a simpler problem* strategy. (Lesson 5–2)

13. Ralph is hanging wallpaper on a wall in the kitchen. The wall is 7 ft by 10 ft, and it contains a doorway that is 6 ft by 3 ft. How much wallpaper will Ralph need? Remember: The area of a rectangle is found by multiplying its lengh times its width.

_____

14. Working separately, 4 students can do 80 math problems in one hour. At this rate, how many math problems can 8 students do in 3 hours?

_____

Name _____ Date _____

Homework Practice

More Algebraic Expressions

Evaluate each expression if $m = 3$ and $n = 15$.

1. $25 - n$

_____

2. $2m - 4$

_____

3. $3n + m$

_____

4. $n - 3$

_____

5. $60 \div n$

_____

6. $2m + n$

_____

7. $2n - m$

_____

8. $6m + 3$

_____

9. $n - 2m$

_____

10. $3m + n$

_____

11. $4n + m$

_____

12. $20 - n$

_____

Evaluate each expression if $a = 2$, $b = 12$, and $c = 8$.

13. $a^2 + 2b$

_____

14. $2c - 3$

_____

15. $b + 3a$

_____

16. $2b + 6$

_____

17. $8a - b$

_____

18. $8c - b$

_____

Spiral Review

Evaluate each expression If $x = 5$ and $y = 2$. (Lesson 5–3)

19. $2x$ _____

20. $5y$ _____

21. $8y$ _____

22. $6y$ _____

23. $3x$ _____

24. $11y$ _____

25. $12x$ _____

26. $9x$ _____

27. $4x$ _____

Name _____ Date _____

Homework Practice

Problem-Solving Investigation: Choose the Best Strategy

Use any strategy shown below to solve each problem.

- Act it out
- Look for a pattern
- Make a table

1. Frank completed 3 passes the first year that he played football, 5 the second year, and 7 the third year. At this rate, how many passes should he expect to complete by his sixth year playing football?

   _____

2. To train for the Math League competition, Janice spent $\frac{1}{2}$ hour each day of the first week reviewing lessons. She added an additional $\frac{1}{2}$ hour per day each week for 4 weeks. What were the total number of hours she spent reviewing during the fourth week?

   _____

3. The table below shows the amount of snow in Maine for 4 weeks during January. How many total inches of snow fell during the month of January?

Week	1	2	3	4
Snow (in.)	21	28	29	22

   _____

Spiral Review

Evaluate each expression if $a = 2$, $b = 3$, and $c = 5$. (Lesson 5-4)

4. $a + 21$ _____

5. $31 + c$ _____

6. $36 - b$ _____

7. $16 \div a$ _____

8. $6a$ _____

9. $21 \div b$ _____

10. $c + a$ _____

11. $c - b$ _____

12. $5b$ _____

5-6

Homework Practice

Function Tables

Complete each function table.

1.

Input (x)	x − 3	Output
5		
8		
4		

2.

Input (x)	3x	Output
4		
2		
9		

3.

Input (x)	2x	Output
4		
3		
12		

4.

Input (x)	x − 5	Output
7		
15		
25		

5. Mrs. Gregg is giving each of her students 3 apple slices for a snack. Find the function rule. Then make a frequency table to find how many students she has if she passes out 54, 60, or 66 slices.

Input (s)		Output
54		
60		
66		

_____

Spiral Review

Use any strategy shown below to solve each problem.

- Solve a simpler problem
- Draw a picture
- Work backward
- Guess and check

6. Pamela is 2 years younger than her sister. Her sister is three times younger than their mother. Their mother is 25 years younger than their grandmother. If their grandmother is 67, how old is Pamela?

_____

7. Tobias is having friends over for dinner. The dinner table is 6 feet by 3 feet. If each friend will need about 2 feet of table space to sit comfortably, what is the maximum number of friends Tobias can invite to dinner?

_____

Name _____ Date _____

Homework Practice

Order of Operations

Find the value of each expression.

1. $2 \times (4 + 7)$ _____

2. $10 \times (6 - 3)$ _____

3. $(15 \div 3) + (9 - 5)$ _____

4. $(66 \div 11) + 3$ _____

5. $13 \times 5 \times (8 - 3)$ _____

6. $(18 - 3) + (9 - 0)$ _____

7. $(27 \div 3) + (38 - 15)$ _____

8. $26 + (6 \times 4)$ _____

9. $8 \div (20 - 16)$ _____

10. $(7 \times 6) \div (9 - 3)$ _____

11. $22 \times (4 \div 4)$ _____

12. $(8 + 32) \times (20 - 10)$ _____

Spiral Review

Complete each function table. (Lesson 5–6)

13. The output value is 7 less than the input value.

Input (x)	x − 7	Output
16	16 − 7	9
17	17 − 7	10
18	18 − 7	11

14. The output value is half of the input value.

Input (x)	x ÷ 2	Output
20	20 ÷ 2	10
22	22 ÷ 2	11
24	24 ÷ 2	12

40

Name _____ Date _____

Homework Practice

Addition and Subtraction Equations

Solve each equation. Check your solution.

1. $n - 14 = 92$ _____

2. $d - 3 = 6$ _____

3. $b - 7 = 3$ _____

4. $r - 41 = 49$ _____

5. $y - 8 = 14$ _____

6. $s - 4 = 1$ _____

7. $19 = d - 62$ _____

8. $m - 7 = 8$ _____

9. $28 = w - 7$ _____

10. $46 = z - 7$ _____

11. $f - 3 = 12$ _____

12. $2 = t - 1$ _____

13. $4 = e + 0$ _____

14. $z + 2 = 6$ _____

15. $529 = g + 300$ _____

16. $c + 200 = 473$ _____

17. $w + 356 = 500$ _____

18. $p + 1 = 7$ _____

19. $e + 211 = 481$ _____

20. $923 = y + 127$ _____

21. $h + 41 = 48$ _____

22. $3 = q + 1$ _____

23. $m + 32 = 75$ _____

24. $d + 4 = 12$ _____

Solve.

25. Leah started with d dollars. After Leah spent $19, she had $13 left. Write a subtraction equation to represent this situation. Then solve the equation to find the amount of money Leah started with.

_____

_____

26. One year Chicago, IL, received 39 inches of snow. That was 10 inches more than the previous year. Write an addition equation to describe the situation. Solve it to find last year's snowfall in inches, s.

_____

_____

Spiral Review

Find the value of each expression. (Lesson 5–7)

27. $(8 + 4) \times 2$ _____

28. $16 - (4 \times 3)$ _____

29. $(5 + 9) - 6$ _____

30. $(12 - 3) \div 3$ _____

Homework Practice

Multiplication Equations

Solve each equation. Check your solution.

1. $6b = 24$ _____

2. $8m = 32$ _____

3. $49 = 7x$ _____

4. $2y = 10$ _____

5. $1t = 12$ _____

6. $63 = 7x$ _____

7. $8a = 24$ _____

8. $2s = 22$ _____

9. $54 = 6y$ _____

Solve.

10. Maria painted flowers on 10 plates and earned $50. Write and solve an equation to find out how much she earned for each plate.

_____

Spiral Review

Solve each equation. Check your solution. (Lesson 6–1)

11. $y - 4 = 2$ _____

12. $4 = 11 - d$ _____

13. $4 - x = 4$ _____

14. $x - 8 = 3$ _____

15. $y - 4 = 7$ _____

16. $8 = 13 - d$ _____

17. $5 = d - 10$ _____

18. $t - 9 = 4$ _____

19. $1 = 10 - d$ _____

Solve.

20. Joe is 3 inches shorter than his brother Jack. Jack is 60 inches tall. Write and solve an equation to find out how tall Joe is.

_____

Name _____ Date _____

Homework Practice

Problem-Solving Strategy: Make a Table

Use the *make a table* strategy to solve.

1. Veronica is saving money to buy a portable digital music player that costs $250. She plans to save $10 the first month and then increase the amount she saves by $5 each month after the first month. How many months will it take her to save $150?

_____

2. A recipe for potato salad calls for one teaspoon of vinegar for every 2 teaspoons of mayonnaise. How many teaspoons of vinegar are needed for 16 teaspoons of mayonnaise?

_____

3. Samson is 6 years old. His mother is 30 years old. How old will Samson be when his mother is exactly 4 times as old as he is?

_____

4. A package of four mechanical pencils comes with two free erasers. If you get a total of 12 free erasers, how many packages of pencils did you buy?

_____

Spiral Review

Solve each equation. Check your solution. (Lesson 6–2)

5. $3s = 27$ _____

6. $49 = 7x$ _____

7. $10 = 5t$ _____

8. $4r = 36$ _____

9. $40 = 8m$ _____

10. $28 = 4n$ _____

11. $72 = 9x$ _____

12. $2s = 4$ _____

Name _____ Date _____

Homework Practice

Geometry: Ordered Pairs

Name the point for each ordered pair.

1. (1, 2) _____

2. (3, 3) _____

3. (5, 8) _____

5. (4, 2) _____

6. (6, 1) _____

7. (2, 1) _____

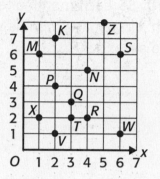

Name the ordered pair for each point.

8. P _____

9. S _____

10. M _____

11. N _____

12. T _____

13. K _____

Spiral Review

Solve. Use the *make a table* strategy. (Lesson 6-3)

14. A recipe for pizza crust calls for 3 cups of flour for every 1 cup of water. How many cups of water are needed for 21 cups of flour?

15. Kristy bought a package of mini muffins for $3. Each package contains 12 mini muffins. If Kristy has 60 mini muffins, how much money did she spend?

16. Andrew is saving to buy a tree house. He saves $2 the first week, $6 the second week, $10 the third week, and so on. How much money will he have saved in 8 weeks?

17. Brett is planning to buy a digital camera. Each month he doubles the amount he saved in the previous month. If he saves $15 the first month, how much money will Brett have saved in 6 months?

Name _____ Date _____

Homework Practice

Algebra and Geometry: Graph Functions

Complete the table. Then graph the ordered pairs.

1. $y = x$

x	1	2	3	4
y	1			

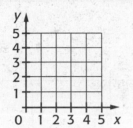

2. $y = x + 2$

x	0	1	2	3
y	2			

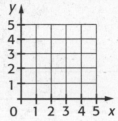

3. $y = 2x$

x	0	1	2	3
y				

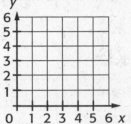

4. $y = 2x - 1$

x	1	2	3	4
y				

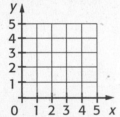

5. $y = x + 1$

x	0	1	2	3
y				

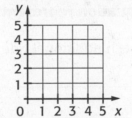

6. $y = x - 1$

x	1	2	3	4
y				

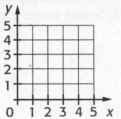

Spiral Review

The graph at the right shows the amount of fuel
a car uses for different distances traveled.

**Use the graph to find the number of gallons of
gasoline used for each distance. (Lesson 6–4)**

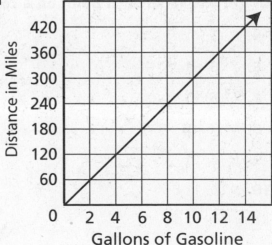

7. 90 mi _____   **8.** 210 mi _____

9. 120 mi _____   **10.** 270 mi _____

11. 330 mi _____   **12.** 300 mi _____

13. 60 mi _____   **14.** 150 mi _____

15. 360 mi _____   **16.** 180 mi _____

17. How many miles per gallon of gasoline
does this car get?

_____

45

Name _____ Date _____

6-6

Homework Practice

Functions and Equations

Complete the table. Write an equation to show the relationship.

1.

Input	x	0	1	2	3	4	5
Output	y	4	10	16	22		

_____

2.

Input	x	0	1	2	3	4	5
Output	y	3	5	7	9		

_____

**Write an equation for the function described in words.
Tell what each variable in the equation represents.**

3. The price equals 5 plus 4 for each additional ride.

_____

4. The total equals 12 minus 6 for each item.

_____

Spiral Review

Graph and label each point on a coordinate grid. (Lesson 6–5)

5. B (3, 4)

6. Y (6, 6)

7. C (5, 2)

8. M (6, 0)

9. R (4, 2)

10. S (3, 1)

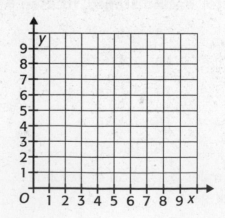

Name _____ Date _____

Homework Practice

Problem-Solving Investigation: Choose the Best Strategy

Use any strategy shown below to solve.

- Logical reasoning
- Work backward
- Guess and check

1. Find the missing term in the pattern below.

1, 1, 2, 3, 5, _____, 13, 21,...

_____

2. Yolanda needs to be home at 4:00 P.M. It takes her 20 minutes to walk home, 20 minutes to say goodbye to her friends, and 10 minutes to organize her books and notebooks at school. What is the latest time she should start getting ready to come home?

_____

Spiral Review **Write an equation for the function described in words. (Lesson 6–6)**

3. The number of baseball cards Tommy has if he gets 5 new cards each week.

_____

4. The cost of belonging to the swim club is $26 per week plus $75 for registration.

_____

5. The dinner cost $13 per person.

_____

6. Jane's height if she is 46 inches tall and grows 1 inch per year.

_____

Name _____ Date _____

Homework Practice

Mean, Median, and Mode

For each data set, find the mean, median, and mode of each set of data.

22, 26, 19, 29, 23, 22, 39, 32

1. Mean _____ 2. Median _____ 3. Mode _____

11, 13, 12, 8, 12,

4. Mean _____ 5. Median _____ 6. Mode _____

28, 33, 39, 44, 52, 52, 11

7. Mean _____ 8. Median _____ 9. Mode _____

$3.25 $4.20 $3.50 $4.25 $3.35 $3.50 $3.50

10. Mean _____ 11. Median _____ 12. Mode _____

Spiral Review

Use any strategy shown below to solve each problem. (Lesson 6-7)

- Make a table
- Guess and check
- Work backward
- Draw a picture

13. Maya has 12 CDs less than Greg. Mohamed has 2 more CDs than Maya. How many CDs do Maya and Mohamed have if Greg has 20 CDs?

_____

14. Winnie is painting a fence. She can paint 3 feet of fencing in an hour. If the fence is 16 feet long, how long will it take her to paint the whole fence?

_____

Name _____ Date _____

Homework Practice

Problem-Solving Investigation: Choose the Best Strategy

Solve.

1. Ryan wants to buy three magazines for $4 each. If he gives the cashier a $20 bill, how much change will he receive?

   _____

2. Tom plays several games of bowling and scores 150, 145, 170, 157, 145, 155, and 163. Find the mean, median, and mode of Tom's bowling scores.

   _____

3. If Jorge practices karate for 45 minutes on Monday, 30 minutes on Tuesday, 35 minutes on Wednesday, and 40 minutes on Thursday, how many minutes should he practice on Friday if he wants to practice for a total of 200 minutes that week?

   _____

4. Lindsey wants to learn how to play some songs on the guitar. She learns one song in her first month of practicing, two more songs in her second month, and three more songs in her third month. If the pattern continues, how many total songs will she know after five months?

   _____

 **Review**

Find the mean, median, and mode of each set of data. (Lesson 7–1)

5. ages of students: 11, 12, 11, 13, 14 _____

6. inches of rain: 3.1, 2.5, 4.2, 3.1, 5.6 _____

7. cost of snacks: $4, $2, $5, $1, $2, $4 _____

7-3

Homework Practice

Line Plots

The line plot below represents the total number of runs scored in each game by Tatiana's softball team this year. Use the information on the line plot to answer the questions.

Number of Runs Scored

1. How many times did the team score 6 runs?

   _____

2. What is the median number of runs scored?

   _____

3. What is the mode of the data?

   _____

```
                              X
                        X     X
                        X     X     X
                        X     X     X     X
         X              X     X     X     X
         X        X     X     X     X     X
    <----+----+----+----+----+----+----+---->
         1    2    3    4    5    6    7
                        Runs
```

4. Find the median, range, and any outliers of the data.

   _____

   _____

Spiral Review

Use any strategy shown below to solve each problem. (Lesson 7-2)

- Guess and check   • Make a table
- Act it out

5. Ellen has two more apples than Nancy. If they have 12 apples altogether, how many apples does each person have?

   _____

6. Four friends raced their bicycles. Beth finished after Mike and before Josh. Matt finished after Beth but before Josh. Who won the race?

   _____

Name _____ Date _____

Homework Practice
Frequency Tables

The Reges family kept a record of the heights of all the children at their family reunion. Here are the results:

51, 45, 48, 51, 51, 50, 45, 47, 45, 45, 46, 49, 46, 50, 49, 45

1. Make a frequency table of the data.

2. Find the median, mode, and range of the data. Identify any outliers.

_____

Height (in.)	Tally	Frequency
45		
46		
47		
48		
49		
50		
51		

Spiral Review

Draw a line plot for the set of data. Then find the median, mode, range, and any outliers of the data shown in the line plot. (Lesson 7-3)

3.

Number of Times on a Plane			
1	2	0	2
1	0	1	3
0	2	0	2
0	3	0	7

Number of Times on a Plane

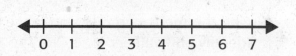

```
0  1  2  3  4  5  6  7
```

_____

Name _____ Date _____

Homework Practice

Scales and Intervals

The table shows the longest rivers in the world.

Longest Rivers in the World (miles)				
2,341	2,590	2,718	2,543	4,160
2,635	2,734	3,362	2,744	2,485
3,395	2,350	4,000	3,964	2,600

1. Choose an appropriate scale and interval size for a frequency table that will represent the data. Then make a frequency table.

_____

Longest Rivers in the World		
Length (mi)	Tally	Frequency

2. Write a sentence or two to describe how the data are distributed among the intervals.

_____

Spiral Review

The table shows the heights of 25 small trees in Lucy's backyard. (Lesson 7-4)

Height of Trees (in.)				
72	73	72	72	72
72	72	73	72	72
74	72	72	74	73
72	72	75	75	73
74	78	73	72	74

Height of Trees		
Height (in.)	Tally	Frequency

3. Make a frequency table of the data.

4. Find the median, mode, and range of the data. Identify any outliers.

_____

_____

Name _____ Date _____

Homework Practice

Bar Graphs

This double-bar graph shows José's best times for three different swimming events over two weeks.

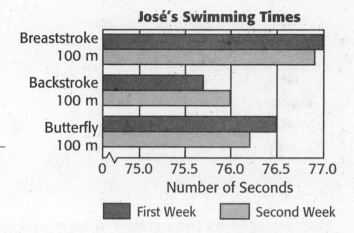

José's Swimming Times

1. In how many seconds did José swim the breaststroke the first week? the second week?

   _____

2. For which events did he improve his time the second week?

   _____

Tina surveyed fifth graders at her school about their favorite sport. The table shows the results of her survey.

Favorite Sport		
Sport	**Ms. Kwan's Class**	**Mr. Brooke's Class**
Soccer	12	14
Basketball	7	5
Track	10	11

3. Make a double bar graph to represent the data. Which sport was the most popular in Mr. Brooke's class? In Ms. Kwan's class?

   _____

Favorite Sport

Spiral Review

4. The heights of Tamika's friends in inches are 60, 52, 54, 48, 52, 59, 55, 60, 51, 50, 52, and 53. Choose an appropriate scale and interval size for a frequency table that will represent the heights. Then make a frequency table. (Lesson 7-5)

5. Write a sentence or two to describe how the data are distributed among the intervals.

Heights of Friends		
Heights (in.)	**Tally**	**Frequency**
	III	
	HHT I	
	I	
	I	

Name _____ Date _____

7-7

Homework Practice

Line Graphs

The table shows the temperature each day at noon for five days.

1. Make a line graph of the data.

Noon Temperatures

Day	Temperature
Monday	1°C
Tuesday	5°C
Wednesday	12°C
Thursday	6°C
Friday	10°C

The table shows the number of boys and girls who used the computer lab after school.

2. Make a double line graph of the data.

Computer Lab After School

Day of the week	Number of boys	Number of girls
Monday	5	10
Tuesday	15	20
Wednesday	25	15
Thursday	10	30
Friday	5	10

Spiral Review

Use the double bar graph to answer Exercises 3–5. (Lesson 7–6)

3. Which game had the most runs scored by the Hawks? _____

4. How many runs did the Rockets score in Game 5? _____

5. In which game did the Rockets score 7 runs? _____

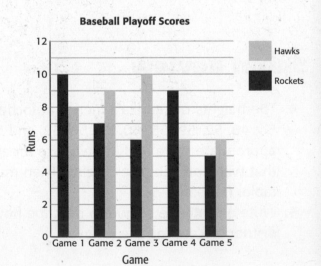

Baseball Playoff Scores

Name _____ Date _____

Homework Practice

Use an Appropriate Graph

Select an appropriate type of display for data gathered about each situation.

1. comparing the costs of four bicycles

_____

2. showing the number of inches of rain each hour during a hurricane

_____

3. the number of boys who attended Camp Green Tree each year from 2006–2008

_____

4. comparing the populations of the largest cities in California

_____

Spiral Review

Refer to the line graph below. It shows Noelle's height in inches each year on her birthday. (Lesson 7–7)

5. In between which two years did she grow the most?

_____

6. In between which years did Noelle grow the least?

_____

7. What is the scale of each axis?

_____

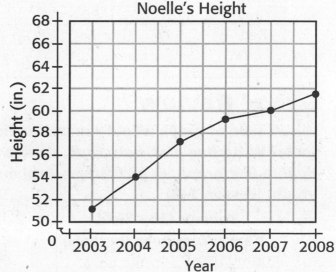

Noelle's Height

7–9

Homework Practice

Problem-Solving Strategy: Make a Graph

Solve by *using a graph*.

1. The table shows the number of hours students spent talking on the phone last month. Find the mode of the data by creating a line plot.

Talking on the Phone (hrs)				
2	2	1	2	0
0	2	3	2	1
1	2	3	2	1
3	2	2	3	1
3	1	1	0	3

_____

2. Kyle surveyed his classmates to see what color shoes they were wearing. How does the number of brown shoes compare to the number of black shoes?

Student Shoe Colors	
Color	Number
Black	11
White	7
Brown	5
Grey	6
Red	1

_____

Spiral Review

Which type of graph would you use to display the data in the table? Write *line plot, bar graph, double bar graph, line graph, double line graph,* or *pictograph*. Explain why. Then make the graph. (Lesson 7–8) _____

3.

Tomato Crop Yield	
Year	Tomatoes (bushels)
2000	40
2001	42
2002	39
2003	46
2004	41
2005	40
2006	38

Name _____ Date _____

Homework Practice

Fractions and Division

Represent each situation using a fraction. Then solve.

1. Three bags of soil are used to fill 4 flowerpots. How many cups of soil does each flowerpot use?

_____

2. Three people equally share five lemon squares. How many lemon squares does each person receive?

_____

3. In science class there are 5 cups of water to be used for the experiments. If six students work on the experiments, how many cups of water does each student receive?

_____

4. Four yards of fabric are used to make five craft projects. How many yards of fabric does each craft project use?

_____

Spiral Review

Which type of graph would you use to display the data in each table? Write bar graph, line graph, or pictograph.

5. Brandon surveyed his classmates to find their favorite sport. Which type of graph should you use to display the data? Which sport is most popular?

Favorite Sports	
Basketball	9
Baseball	6
Football	18
Soccer	7
Lacrosse	3

6. The following table shows the height of five students. Which type of graph should you use? What is the median of these heights?

Student Heights	
Name	Height (in.)
Jennifer	58
Troy	55
Bianca	49
Rosa	50
Anna	42

Name _____ Date _____

Homework Practice

Improper Fractions

Write each improper fraction as a mixed number.

1. $\frac{11}{6}$ _____

2. $\frac{13}{4}$ _____

3. $\frac{41}{7}$ _____

4. $\frac{19}{4}$ _____

5. $\frac{5}{2}$ _____

6. $\frac{38}{5}$ _____

7. $\frac{9}{2}$ _____

8. $\frac{14}{3}$ _____

9. $\frac{39}{8}$ _____

10. $\frac{25}{6}$ _____

11. $\frac{22}{5}$ _____

12. $\frac{17}{4}$ _____

13. $\frac{80}{9}$ _____

14. $\frac{13}{10}$ _____

15. $\frac{67}{7}$ _____

16. $\frac{71}{8}$ _____

17. $\frac{8}{3}$ _____

18. $\frac{14}{5}$ _____

19. $\frac{28}{3}$ _____

20. $\frac{61}{7}$ _____

21. $\frac{13}{6}$ _____

Spiral Review

Represent each situation using a fraction. Then solve. (Lesson 8–1)

22. Eight people equally share 3 pizzas. How much pizza does each person recieve?

_____

23. In art class, there are 5 sheets of drawing paper for 9 people. How much paper will each person receive?

_____

24. Five gallons of punch fill 3 punch bowls equally. How much punch will be in each punch bowl?

_____

Name _____ Date _____

Homework Practice

Problem-Solving Strategy: Use Logical Reasoning

Solve. Use logical reasoning.

1. For a science lesson, Mr. Miller asked his students to each bring in a leaf or a pinecone. The students brought in 21 leaves and pinecones in all. There were 5 more leaves than pinecones. How many pinecones did the students bring in?

   _____

2. Jenna can swim one lap in the pool in 36 seconds. How long will it take her to swim 3 laps?

   _____

3. There are 16 ounces in 1 pound. How many ounces are there in 3 pounds?

   _____

4. Marcus is practicing for the basketball team. The chart below shows the number of minutes he has practiced for each of the last 4 days. If the pattern continues, how many minutes will he practice on the fifth day?

   _____

Day	Time (in minutes)
One	30
Two	45
Three	60
Four	75
Five	_____

Spiral Review

Write each improper fraction as a mixed number. (Lesson 8–2)

5. $\frac{7}{2}$ _____

6. $\frac{5}{3}$ _____

7. $\frac{12}{5}$ _____

8. $\frac{15}{2}$ _____

9. $\frac{18}{7}$ _____

10. $\frac{9}{4}$ _____

Name _____ Date _____

Homework Practice

Mixed Numbers

Write each mixed number as an improper fraction.

1. $2\frac{3}{4}$ _____

2. $5\frac{1}{6}$ _____

3. $8\frac{1}{2}$ _____

4. $3\frac{2}{3}$ _____

5. $7\frac{2}{5}$ _____

6. $1\frac{9}{10}$ _____

7. $4\frac{7}{8}$ _____

8. $6\frac{5}{7}$ _____

9. $1\frac{8}{9}$ _____

10. $3\frac{12}{17}$ _____

11. $2\frac{1}{10}$ _____

12. $5\frac{5}{13}$ _____

13. $1\frac{1}{2}$ _____

14. $7\frac{1}{3}$ _____

15. 3 _____

16. $3\frac{1}{2}$ _____

17. $4\frac{2}{3}$ _____

18. 8 _____

19. $2\frac{3}{5}$ _____

20. $5\frac{3}{4}$ _____

21. $2\frac{5}{8}$ _____

22. $1\frac{29}{35}$ _____

23. $6\frac{1}{3}$ _____

24. $5\frac{1}{2}$ _____

25. $3\frac{7}{10}$ _____

26. $4\frac{1}{2}$ _____

27. $4\frac{1}{10}$ _____

28. $5\frac{2}{5}$ _____

29. $8\frac{3}{4}$ _____

30. $2\frac{3}{5}$ _____

Spiral Review

Solve. Use logical reasoning.

31. A shipment of boxes weighs 40 pounds. There are 8 boxes and each weighs the same number of pounds. How much does each box weigh?

32. Mrs. Cooper's fifth-grade class has 11 more girls than boys. There are 35 students in all. How many girls are there?

_____

_____

8-5

Homework Practice

Fractions on a Number Line

Replace each ◯ with < or > to make a true statement.

1. $\frac{2}{3}$ ◯ $\frac{5}{3}$

2. $3\frac{3}{8}$ ◯ $\frac{28}{8}$

3. $\frac{3}{7}$ ◯ $\frac{2}{7}$

4. $\frac{11}{9}$ ◯ $1\frac{3}{9}$

5. $1\frac{2}{5}$ ◯ $\frac{8}{5}$

6. $\frac{16}{7}$ ◯ $2\frac{5}{7}$

7. $\frac{9}{4}$ ◯ $1\frac{3}{4}$

8. $\frac{13}{10}$ ◯ $1\frac{1}{10}$

9. $\frac{13}{8}$ ◯ $2\frac{1}{8}$

Write the fraction or mixed number that is represented by each point.

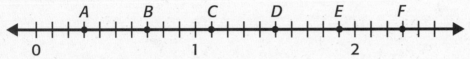

10. A _____

11. B _____

12. C _____

13. D _____

14. E _____

15. F _____

Spiral Review

Write each mixed number as an improper fraction. (Lesson 8–4)

16. $2\frac{3}{5}$ _____

17. $5\frac{1}{10}$ _____

18. $4\frac{5}{8}$ _____

19. $11\frac{4}{5}$ _____

20. $6\frac{1}{7}$ _____

21. $7\frac{2}{9}$ _____

Name _____ Date _____

Homework Practice

Round Fractions

Round each number to 0, $\frac{1}{2}$, or 1.

1. $\frac{1}{12}$ _____

2. $\frac{5}{11}$ _____

3. $\frac{3}{10}$ _____

4. $\frac{8}{12}$ _____

5. $\frac{2}{9}$ _____

6. $\frac{14}{16}$ _____

7. $\frac{6}{16}$ _____

8. $\frac{7}{12}$ _____

9. $\frac{3}{8}$ _____

Solve.

10. Your basement has an $8\frac{3}{12}$ foot ceiling. To the nearest half foot, how tall is the tallest cabinet that can fit in the basement?

_____

11. Alice is giving a book as a gift that is $8\frac{3}{8}$ inches long and $6\frac{1}{12}$ inches wide. Will the book fit in a box that is $8\frac{1}{2}$ inches long and $6\frac{1}{2}$ inches wide or in a box that is 8 inches long and 6 inches wide?

_____

Spiral Review

Replace $\bigcirc$ with < or > to make a true statement. (Lesson 8-5).

12. $\frac{3}{4}\bigcirc\frac{1}{4}$

13. $\frac{4}{7}\bigcirc\frac{5}{7}$

14. $2\frac{1}{9}\bigcirc 1\frac{2}{9}$

15. $1\frac{2}{3}\bigcirc 2\frac{1}{3}$

16. $\frac{9}{6}\bigcirc\frac{5}{6}$

17. $3\frac{1}{12}\bigcirc 2\frac{11}{12}$

Name _____ Date _____

Homework Practice

Problem-Solving Investigation: Choose the Best Strategy

Use any strategy shown below to solve.

- Guess and check
- Work backward
- Solve a simpler problem
- Make a table
- Use logical reasoning
- Act it out

1. Olivia bought a ring for $\frac{1}{2}$ off the regular price. If she paid $50, what was the regular price?

_____

2. Mrs. Jones told the class that $\frac{1}{3}$ of them scored 90 or above on the math test. Another $\frac{1}{3}$ of them had a passing score. What fraction of the class failed?

_____

3. At a park, a picnic shelter covers $\frac{1}{4}$ of an acre and a playground covers $\frac{5}{8}$ of an acre. How much area is covered by both the picnic shelter and the playground?

_____

4. Of the 300 students at school, 110 are in the chorus and 150 are in the band. Of these students, 50 are in both chorus and the band. How many students are neither in the chorus nor the band?

_____

Spiral Review

Round each fraction to 0, $\frac{1}{2}$, or 1.

5. $\frac{1}{7}$

6. $\frac{7}{8}$

7. $\frac{2}{10}$

8. $\frac{5}{6}$

9. $\frac{5}{9}$

10. $\frac{4}{10}$

Name _____ Date _____

Homework Practice

Common Factors

Find the common factors of each set of numbers.

1. 1, 4, 8, 32 _____

2. 1, 3, 6, 12, 24 _____

Find the GCF of each set of numbers.

3. 5, 45 _____

4. 6, 42 _____

5. 12, 24, 60 _____

6. 4, 16, 32 _____

7. 15, 30, 60 _____

8. 9, 18, 27 _____

Solve.

9. Janice has three CD storage cases that can hold 18, 36, and 72 CDs. The cases have sections holding the same number of CDs. What is the greatest number of CDs in a section?

   _____

10. Packages of cheese are sold in sealed containers that have sections holding the same number of slices. The containers can hold 6, 12, and 24 sections. What is the greatest number of sections in each container? _____

Spiral Review

Use any strategy to solve each problem. (Lesson 8–7)

11. A movie starts at 6:45 P.M. and lasts 1 hour and 35 minutes. At what time does the movie end? _____

12. What two positive integers have a sum of 15 and a product of 56?

    _____

Name _____ Date _____

Homework Practice

Prime and Composite Numbers

Tell whether each number is *prime* or *composite*.

1. 75

_____

2. 61

_____

3. 96

_____

4. 48

_____

5. 29

_____

6. 95

_____

7. 68

_____

8. 54

_____

9. 171

_____

10. 143

_____

11. 117

_____

12. 209

_____

Problem Solving.
Solve.

13. A board is 24 inches long. Find all the whole-number lengths into which it can be evenly divided.

_____

14. A ribbon is 36 inches long. Find all the whole-number lengths into which it can be evenly divided.

_____

Spiral Review

Find the GCF of each set of numbers. (Lesson 9–1)

15. 6, 18

_____

16. 16, 30

_____

17. 14, 28

_____

18. 27, 54

_____

19. 8, 12

_____

20. 49, 63

_____

Name _____ Date _____

Homework Practice

Equivalent Fractions

Write two fractions that are equivalent to each fraction.

1. $\dfrac{2}{5}$ _____

2. $\dfrac{6}{18}$ _____

3. $\dfrac{5}{10}$ _____

4. $\dfrac{3}{12}$ _____

5. $\dfrac{21}{35}$ _____

6. $\dfrac{6}{8}$ _____

7. $\dfrac{8}{20}$ _____

8. $\dfrac{3}{9}$ _____

9. $\dfrac{12}{15}$ _____

10. $\dfrac{6}{24}$ _____

11. $\dfrac{12}{20}$ _____

12. $\dfrac{8}{10}$ _____

Algebra Find the number for $\square$ that makes the fractions equivalent.

13. $\dfrac{4}{5} = \dfrac{\square}{15}$

14. $\dfrac{14}{16} = \dfrac{\square}{8}$

15. $\dfrac{3}{\square} = \dfrac{12}{28}$

16. $\dfrac{2}{6} = \dfrac{\square}{3}$

17. $\dfrac{\square}{16} = \dfrac{3}{4}$

18. $\dfrac{3}{4} = \dfrac{6}{\square}$

19. $\dfrac{14}{42} = \dfrac{\square}{3}$

20. $\dfrac{5}{\square} = \dfrac{15}{27}$

21. $\dfrac{9}{30} = \dfrac{3}{\square}$

Are the fractions equivalent? Write yes or no.

22. $\dfrac{3}{4} = \dfrac{6}{8}$ _____

23. $\dfrac{3}{8} = \dfrac{7}{16}$ _____

24. $\dfrac{5}{9} = \dfrac{15}{27}$ _____

25. $\dfrac{2}{3} = \dfrac{4}{5}$ _____

26. $\dfrac{5}{7} = \dfrac{15}{21}$ _____

27. $\dfrac{10}{13} = \dfrac{7}{14}$ _____

Spiral Review

Tell whether each number is prime or composite. (Lesson 9–2)

28. 14 _____

29. 33 _____

30. 29 _____

31. 47 _____

32. 18 _____

33. 28 _____

Name _____ Date _____

Homework Practice

Simplest Form

Replace each □ with a number so the fractions are in simplest form.

1. $\dfrac{6}{16} = \dfrac{3}{\square}$

2. $\dfrac{5}{15} = \dfrac{1}{\square}$

3. $\dfrac{10}{35} = \dfrac{\square}{7}$

4. $\dfrac{20}{25} = \dfrac{4}{\square}$

Write each fraction in simplest form. If the fraction is already in simplest form, write *simplified*.

5. $\dfrac{2}{4}$ _____

6. $\dfrac{1}{3}$ _____

7. $\dfrac{12}{16}$ _____

8. $\dfrac{9}{10}$ _____

9. $\dfrac{4}{15}$ _____

10. $\dfrac{2}{10}$ _____

Spiral Review

Write two fractions that are equivalent to each fraction.   (Lesson 9–3)

11. $\dfrac{3}{4}$ _____

12. $\dfrac{1}{3}$ _____

13. $\dfrac{6}{8}$ _____

14. $\dfrac{2}{5}$ _____

15. $\dfrac{1}{6}$ _____

16. $\dfrac{1}{2}$ _____

Name _____ Date _____

Homework Practice

Decimals and Fractions

Write each decimal as a fraction in simplest form.

1. 0.2 _____

2. 6.12 _____

3. 0.375 _____

4. 0.32 _____

5. 0.125 _____

6. The newspaper reported that it rained 2.20 inches last month. Express this amount as a mixed number in simplest form.

_____

Write each decimal as a mixed number in simplest form.

7. 6.3 _____

8. 32.50 _____

9. 40.330 _____

10. 24.500 _____

Spiral Review

Write each fraction in simplest form. If the fraction is already in simplest form, write *simplified*.   (Lesson 9–4)

11. $\frac{12}{27}$ _____

12. $\frac{7}{9}$ _____

13. $\frac{24}{64}$ _____

14. $\frac{17}{41}$ _____

15. $\frac{15}{35}$ _____

16. $\frac{38}{42}$ _____

Name _____ Date _____

Homework Practice

Problem-Solving Strategy: Look for a Pattern

Solve. Use the *look for a pattern* strategy.

1. Dennis planted his garden so that the tallest plants are at one end. The height of the plants gradually decreases until the shortest plants are at the other end. Frost is predicted for tonight, so he wants to place cones over the plants to protect them. The cones for the smallest plants are 12 inches tall. If each row of cones increases in height by 4 inches, how tall are the tallest cones if there are six different types of plants?

_____

2. Draw the next two figures in the pattern below.

_____

3. Jason built a 4.6-foot tower of snow in his yard. The weather became warmer. After one hour, the tower was 4.4 feet tall, and after two hours, it was 4.2 feet tall. If weather conditions stay the same, how tall will the tower be after 10 hours?

_____

4. Bonita drew stars to cut out for a school program. She first cut a 4-centimeter star. Then she cut a 3.5-centimeter star, followed by a 3-centimeter star. If she follows this pattern, what size will the next star be?

_____

 **Review**

Write each decimal as a fraction in simplest form.   (Lesson 9–5)

5. 0.8 _____

6. 0.04 _____

7. 0.25 _____

8. 0.3 _____

Name _____ Date _____

Homework Practice

Multiples

Identify the first three common multiples of each set of numbers.

1. 3, 15 _____

2. 2, 8, 12 _____

3. 6, 9, 10 _____

4. 3, 6, 18 _____

Find the LCM of each set of numbers.

5. 2, 5 _____

6. 6, 15 _____

7. 4, 16, 32 _____

8. 2, 16, 20 _____

Solve.

9. Find the two missing common multiples from the list of common multiples for 4 and 12.

48, 60, _____, 84, _____, 108, 120

10. For the drama club picture, the students must line up in even rows. Describe the arrangements for the least number of people needed to be able to line up in rows of 5 or 6.

_____

Spiral Review

Solve. Use the *look for a pattern* strategy.   (Lesson 9–6)

11. Mike is filling a bucket. He measures the depth of the water in inches every minute. Here are his measurements: 1.1, 2.3, 3.5, 4.7. If this pattern continues, how deep will the water be the next time he measures?

_____

12. Jack is increasing the distance he runs each week over time. During the first four weeks he ran 3, 4.5, 6, and 7.5 miles. Based on this pattern, how far will he run during the fifth week?

_____

Name _____ Date _____

Homework Practice

Problem-Solving Investigation: Choose the Best Strategy

Use any strategy shown below to solve.

- Guess and check.
- Act it out.
- Make a table.

1. Janet spent a total of $60 on summer clothes. At least 2 of the pairs of shorts she bought cost $10 each. She bought some T-shirts for $5 each. She also bought some sandals for $10. How many of each clothing item did Janet purchase?

   _____

2. Marge went on a trip to New York City and spent a total of $200 going to the theatre. She purchased 4 student tickets for Broadway plays that cost $25 each and five discount tickets. How much did each discount ticket cost?

   _____

3. A radio station is giving every third caller a T-shirt and every tenth caller a ceramic mug. Which caller will be the first to receive both prizes?

   _____

Spiral Review

Identify the first three common multiples of each set of numbers. (Lesson 9–7)

4. 2, 5 _____   5. 6, 9, 18 _____

6. 3, 6, 10 _____   7. 5, 7, 15 _____

Find the LCM of each set of numbers.

8. 8, 16 _____   9. 7, 10 _____   10. 6, 12, 24 _____

9–9

Homework Practice

Compare Fractions

Replace each _____ with <, >, or = to make a true statement.

1. $\dfrac{1}{2}$ _____ $\dfrac{3}{5}$

2. $\dfrac{3}{4}$ _____ $\dfrac{7}{8}$

3. $\dfrac{7}{8}$ _____ $\dfrac{7}{9}$

4. $\dfrac{5}{12}$ _____ $\dfrac{3}{8}$

5. $8\dfrac{1}{8}$ _____ $8\dfrac{2}{3}$

6. $5\dfrac{1}{3}$ _____ $5\dfrac{7}{8}$

Solve.

7. Andrea is using three frames, each with a different width, to frame her photographs. The sizes in inches are $8\dfrac{1}{2}$, $8\dfrac{1}{3}$, $8\dfrac{5}{6}$. She has decided to put the smallest in the center when she hangs them beside each other on the wall. What size frame will be in the center?

_____

Spiral Review

Use any strategy shown below to solve each problem. (Lesson 9–8)

- Act it out.
- Look for a pattern.
- Guess and Check.
- Solve a simpler problem.

8. For a yearbook picture, the 20 baseball team members must line up with an equal number of people in each row. Describe the possible arrangements in which the players could be lined up.

_____

9. Mark needs to mow the grass, trim the hedges, and sweep the front steps before his mother gets home from work. How many different ways can Mark order these activities?

_____

Name _____ Date _____

Homework Practice

Add Like Fractions

Add. Write each sum in simplest form.

1. $\dfrac{2}{5} + \dfrac{8}{5} =$ _____

2. $\dfrac{5}{9} + \dfrac{1}{9} =$ _____

3. $\dfrac{6}{8} + \dfrac{5}{8} =$ _____

4. $\dfrac{3}{4} + \dfrac{2}{4} =$ _____

5. $\dfrac{9}{9} + \dfrac{3}{9} =$ _____

6. $\dfrac{7}{8} + \dfrac{2}{8} =$ _____

7. $\dfrac{1}{2} + \dfrac{2}{2} =$ _____

8. $\dfrac{4}{5} + \dfrac{3}{5} =$ _____

9. $\dfrac{12}{15} + \dfrac{3}{15} =$ _____

10. $\dfrac{6}{7} + \dfrac{1}{7} =$ _____

11. Jasmine ate $\dfrac{3}{8}$ of a pizza. Manny ate $\dfrac{2}{8}$ of the same pizza. How much pizza did they eat altogether? Write a fraction in simplest form.

12. Deanna walked $\dfrac{4}{15}$ of a mile. Abi walked $\dfrac{5}{15}$ of a mile. How far did they walk altogether? Write as a fraction in simplest form.

_____

_____

Spiral Review

Replace each ◯ with <, >, or = to make a true statement.
(Lesson 9–9)

13. $\dfrac{1}{4}$ ◯ $\dfrac{3}{8}$

14. $\dfrac{2}{3}$ ◯ $\dfrac{6}{9}$

15. $\dfrac{1}{2}$ ◯ $\dfrac{5}{9}$

16. $\dfrac{1}{5}$ ◯ $\dfrac{2}{7}$

17. $\dfrac{3}{4}$ ◯ $\dfrac{5}{8}$

18. $\dfrac{7}{12}$ ◯ $\dfrac{6}{13}$

Name _____ Date _____

Homework Practice

Subtract Like Fractions

Subtract. Write each difference in simplest form.

1. $\dfrac{8}{5} - \dfrac{2}{5} = $ _____

2. $\dfrac{5}{9} - \dfrac{1}{9} = $ _____

3. $\dfrac{6}{8} - \dfrac{5}{8} = $ _____

4. $\dfrac{3}{4} - \dfrac{2}{4} = $ _____

5. $\dfrac{9}{9} - \dfrac{3}{9} = $ _____

6. $\dfrac{7}{8} - \dfrac{2}{8} = $ _____

7. $\dfrac{2}{2} - \dfrac{1}{2} = $ _____

8. $\dfrac{4}{5} - \dfrac{3}{5} = $ _____

9. $\dfrac{12}{15} - \dfrac{3}{15} = $ _____

10. $\dfrac{6}{7} - \dfrac{1}{7} = $ _____

Spiral Review

Add. Write each sum in simplest form. (Lesson 10–1)

11. $\dfrac{1}{9} + \dfrac{5}{9} = $ _____

12. $\dfrac{4}{6} + \dfrac{1}{6} = $ _____

13. $\dfrac{2}{3} + \dfrac{1}{3} = $ _____

14. $\dfrac{7}{8} + \dfrac{2}{8} = $ _____

15. $\dfrac{2}{10} + \dfrac{1}{10} = $ _____

16. $\dfrac{1}{3} + \dfrac{6}{3} = $ _____

17. $\dfrac{5}{8} + \dfrac{3}{8} = $ _____

18. $\dfrac{5}{15} + \dfrac{5}{15} = $ _____

19. $\dfrac{7}{8} + \dfrac{1}{8} = $ _____

20. $\dfrac{2}{8} + \dfrac{5}{8} = $ _____

21. $\dfrac{5}{8} + \dfrac{11}{8} = $ _____

22. $\dfrac{6}{7} + \dfrac{2}{7} = $ _____

Name _____ Date _____

Homework Practice

Add Unlike Fractions

Add. Write your answer in simplest form.

1. $\dfrac{2}{3}$
   $+\dfrac{3}{5}$

2. $\dfrac{2}{3}$
   $+\dfrac{5}{9}$

3. $\dfrac{3}{4}$
   $+\dfrac{5}{8}$

4. $\dfrac{2}{7}$
   $+\dfrac{5}{14}$

5. $\dfrac{1}{2}$
   $+\dfrac{5}{6}$

6. $\dfrac{11}{12}$
   $+\dfrac{3}{4}$

7. $\dfrac{5}{12}$
   $+\dfrac{1}{4}$

8. $\dfrac{7}{15}$
   $+\dfrac{1}{6}$

9. $\dfrac{8}{9}$
   $+\dfrac{2}{3}$

10. $\dfrac{5}{6}$
    $+\dfrac{3}{8}$

11. $\dfrac{7}{15}$
    $+\dfrac{1}{3}$

12. $\dfrac{3}{4}$
    $+\dfrac{3}{10}$

13. $\dfrac{2}{9}$
    $+\dfrac{5}{6}$

14. $\dfrac{4}{5}$
    $+\dfrac{3}{4}$

15. $\dfrac{11}{12}$
    $+\dfrac{7}{8}$

16. $\dfrac{7}{10}$
    $+\dfrac{1}{6}$

17. $\dfrac{7}{8}$
    $+\dfrac{2}{3}$

18. $\dfrac{9}{10}$
    $+\dfrac{9}{15}$

19. $\dfrac{2}{5}+\dfrac{7}{10}=$ _____

20. $\dfrac{5}{6}+\dfrac{4}{9}=$ _____

21. $\dfrac{2}{3}+\dfrac{1}{4}=$ _____

22. $\dfrac{7}{10}+\dfrac{1}{5}=$ _____

23. $\dfrac{3}{4}+\dfrac{1}{3}=$ _____

24. $\dfrac{5}{6}+\dfrac{2}{9}=$ _____

25. $\dfrac{2}{5}+\dfrac{3}{10}=$ _____

26. $\dfrac{3}{4}+\dfrac{2}{3}=$ _____

27. $\dfrac{3}{10}+\dfrac{3}{4}=$ _____

Spiral Review

Solve.

28. Cathy spent $\dfrac{2}{5}$ of an hour on her French assignment and $\dfrac{4}{5}$ of an hour on her English report. How much more time did she spend on her English report than her French assignment? Write your answer in simplest form.

29. On saturday, Jason spent $\dfrac{9}{10}$ of his time skateboarding and $\dfrac{1}{10}$ of his time reading. How much more time did Jason spend skateboarding than reading?

_____

_____

10-4

Homework Practice

Subtract Unlike Fractions

Subtract. Write your answer in simplest form.

1. $\dfrac{2}{3}$
   $-\dfrac{3}{5}$

2. $\dfrac{2}{3}$
   $-\dfrac{9}{9}$

3. $\dfrac{3}{4}$
   $-\dfrac{5}{8}$

4. $\dfrac{5}{7}$
   $-\dfrac{5}{14}$

5. $\dfrac{1}{2}$
   $-\dfrac{1}{6}$

6. $\dfrac{11}{12}$
   $-\dfrac{3}{4}$

7. $\dfrac{5}{12}$
   $-\dfrac{1}{4}$

8. $\dfrac{7}{15}$
   $-\dfrac{1}{6}$

9. $\dfrac{8}{9}$
   $-\dfrac{2}{3}$

10. $\dfrac{5}{6}$
    $-\dfrac{3}{8}$

11. $\dfrac{7}{15}$
    $-\dfrac{1}{3}$

12. $\dfrac{3}{4}$
    $-\dfrac{4}{10}$

13. $\dfrac{8}{9}$
    $-\dfrac{5}{6}$

14. $\dfrac{4}{5}$
    $-\dfrac{3}{4}$

15. $\dfrac{11}{12}$
    $-\dfrac{7}{8}$

16. $\dfrac{7}{10}$
    $-\dfrac{1}{6}$

17. $\dfrac{7}{4}$
    $-\dfrac{5}{8}$

18. $\dfrac{9}{10}$
    $-\dfrac{9}{15}$

19. $\dfrac{4}{5} - \dfrac{7}{10} =$ _____

20. $\dfrac{5}{6} - \dfrac{4}{9} =$ _____

21. $\dfrac{2}{3} - \dfrac{1}{4} =$ _____

22. $\dfrac{7}{10} - \dfrac{1}{5} =$ _____

23. $\dfrac{3}{4} - \dfrac{1}{3} =$ _____

24. $\dfrac{5}{6} - \dfrac{2}{9} =$ _____

25. $\dfrac{2}{5} - \dfrac{1}{6} =$ _____

26. $\dfrac{3}{4} - \dfrac{2}{3} =$ _____

27. $\dfrac{9}{10} - \dfrac{3}{4} =$ _____

Spiral Review

Solve.

28. Clifton spent $\dfrac{2}{3}$ hour practicing guitar. He spent $\dfrac{1}{6}$ hour changing the strings on his guitar. How much time did he spend practicing and changing the strings?

_____

29. In the new den, $\dfrac{1}{6}$ of the walls will be made of glass blocks, and $\dfrac{1}{8}$ will be covered with tile. What fraction of the room will be covered with glass blocks and tile?

_____

Name _____ Date _____

Homework Practice

Problem-Solving Strategy: Determine Reasonable Answers

Solve. Determine which answer is reasonable.

1. Marci found $1.42 in her coat pocket. She had $4.85 in her backpack. Is $5.50, $6.50, or $7.50 a more reasonable estimate for how much money she had altogether?

   _____

2. James and a friend picked strawberries. James picked $4\frac{3}{5}$ pounds, and his friend picked $5\frac{4}{5}$ pounds. Which is a more reasonable estimate for how many pounds they picked altogether: 10 pounds, 11 pounds, or 12 pounds?

   _____

3. After school, Philipe spent $1\frac{3}{4}$ hour at baseball practice, $2\frac{1}{4}$ hour on homework, and $\frac{1}{4}$ hour getting ready for bed. Which is a more reasonable estimate for how long he spent on his activities: 3 hours, 4 hours, or 5 hours?

   _____

4. Lynn went shopping at a local store. She bought 5 CDs, for $15.99 each, some candy for $1.79, and gloves for $5.89. Is $85, $88, or $90 a more reasonable estimate for how much money she spent altogether?

   _____

Spiral Review

Subtract. Write each difference in simplest form. (Lesson 10–4)

5. $\frac{3}{6} - \frac{2}{12} =$ _____

6. $\frac{5}{5} - \frac{9}{15} =$ _____

7. $\frac{1}{2} - \frac{2}{8} =$ _____

8. $\frac{3}{4} - \frac{3}{8} =$ _____

9. $\frac{10}{12} - \frac{1}{2} =$ _____

10. $\frac{8}{9} - \frac{2}{3} =$ _____

11. $\frac{4}{5} - \frac{12}{20} =$ _____

12. $\frac{2}{3} - \frac{7}{15} =$ _____

Name _____ Date _____

Homework Practice

Estimate Sums and Differences

Estimate.

1. $4\frac{1}{3} + \frac{8}{9}$ _____

2. $7\frac{1}{6} + \frac{8}{15}$ _____

3. $\frac{9}{10} + 3\frac{2}{3}$ _____

4. $8\frac{7}{8} - 1\frac{6}{9}$ _____

5. $1\frac{2}{10} + 3\frac{1}{9}$ _____

6. $7\frac{1}{3} + 7\frac{1}{8}$ _____

7. $3\frac{5}{8} + 6\frac{3}{5}$ _____

8. $\frac{8}{15} + 2\frac{5}{9}$ _____

9. $6\frac{7}{8} - \frac{4}{7}$ _____

10. $10\frac{7}{8} - \frac{5}{9}$ _____

Spiral Review

Solve. Determine which answer is reasonable. (Lesson 10–5)

11. A store sells 12 pounds of apples. Of those, $5\frac{1}{2}$ pounds are green apples and $2\frac{1}{4}$ are golden. Which is a more reasonable estimate for how many more pounds of green apples than golden apples were sold: 3 pounds, 4 pounds, or 5 pounds?

_____

12. Kelly has $92.63 in the bank. She wants a jacket for $91.00, but must keep at least $25 in the bank. Is $20, $25, or $30 a more reasonable estimate for how much more money she needs?

_____

Name _____ Date _____

Homework Practice

Add Mixed Numbers

Add. Write each sum in simplest form.

1. $7\frac{15}{16} - 2\frac{11}{16} =$ _____

2. $11\frac{8}{10} + 4\frac{3}{10} =$ _____

3. $12\frac{1}{3} + 9\frac{1}{3} =$ _____

4. $18\frac{1}{6} + 9\frac{5}{6} =$ _____

5. $9\frac{2}{12} + 5\frac{1}{12} =$ _____

6. $16\frac{1}{3} + 7\frac{7}{10} =$ _____

7. $34\frac{11}{20} + 15\frac{1}{5} =$ _____

8. $64\frac{3}{4} + 37\frac{11}{12} =$ _____

9. $51\frac{2}{5} + 25\frac{3}{4} =$ _____

10. $46\frac{1}{4} + 27\frac{3}{4} =$ _____

11. $82\frac{4}{5} + 62\frac{2}{5} =$ _____

12. $23\frac{1}{8} + 15\frac{2}{5} =$ _____

13. $16\frac{1}{4} + 7\frac{11}{12} =$ _____

14. $35\frac{7}{8} + 21\frac{4}{16} =$ _____

15. $97\frac{3}{5} + 87\frac{12}{15} =$ _____

16. $\begin{aligned}6\frac{11}{12}\\+\ 4\frac{5}{12}\\\hline\end{aligned}$

17. $\begin{aligned}11\frac{2}{5}\\+\ 3\frac{2}{5}\\\hline\end{aligned}$

18. $\begin{aligned}14\frac{14}{16}\\+\ 5\frac{6}{8}\\\hline\end{aligned}$

19. $\begin{aligned}15\frac{1}{7}\\+\ 6\frac{1}{4}\\\hline\end{aligned}$

Spiral Review

Estimate. (Lesson 10–6)

20. $\frac{4}{6} + 1\frac{5}{6} =$ _____

21. $6\frac{9}{10} - 1\frac{2}{10} =$ _____

22. $19\frac{1}{10} + 5\frac{9}{10} =$ _____

23. $8\frac{11}{12} - 7\frac{1}{12} =$ _____

Name _____ Date _____

Homework Practice

Subtract Mixed Numbers

Subtract. Write each difference in simplest form.

1. $2\frac{3}{4} - 1\frac{5}{8} =$ _____

2. $3\frac{2}{3} - 2\frac{1}{6} =$ _____

3. $3\frac{7}{12} - 1\frac{5}{12} =$ _____

4. $7\frac{3}{4} - 3\frac{7}{12} =$ _____

5. $4\frac{7}{9} - 2\frac{4}{9} =$ _____

6. $6\frac{3}{4} - 4\frac{1}{4} =$ _____

7. $3\frac{1}{2} - 1\frac{1}{2} =$ _____

8. $4\frac{1}{2} - 2\frac{3}{8} =$ _____

9. $7\frac{1}{2} - 5\frac{4}{6} =$ _____

10. $12\frac{5}{8} - 4\frac{3}{8} =$ _____

11. $7\frac{9}{10} - \frac{4}{5} =$ _____

12. $13\frac{4}{5} - 4\frac{2}{5} =$ _____

13. $7\frac{20}{24} - 3\frac{6}{24} =$ _____

14. $12\frac{1}{2} - 4\frac{3}{10} =$ _____

15. $11\frac{3}{8} - 6\frac{1}{8} =$ _____

16. $14\frac{6}{10} - 6\frac{5}{10} =$ _____

17. $15\frac{3}{4} - 9\frac{2}{8} =$ _____

18. $17\frac{9}{10} - 8\frac{3}{10} =$ _____

Spiral Review

Add. Write each sum in simplest form. (Lesson 10–7)

19. $4\frac{3}{4} + 2\frac{3}{4} =$ _____

20. $5\frac{4}{9} + 4\frac{3}{9} =$ _____

21. $6\frac{5}{12} + 3\frac{1}{12} =$ _____

22. $8\frac{3}{7} + 5\frac{4}{7} =$ _____

Name _____ Date _____

Homework Practice

Problem-Solving Investigation: Choose the Best Strategy

Use any strategy to solve each problem.

1. Describe the pattern below. Then, find the missing number.
   50, 500, ___, 50,000.

   _____

2. Melinda's mother is four times as old as Melinda. In 16 years, her mother will be twice her age. How old is Melinda now?

   _____

3. Ginny has a piece of fabric 20 yards long. How many cuts will she make if she cuts the fabric into sections that are 2 yards long?

   _____

Spiral Review

Solve. Write each answer in simplest form. (Lesson 10–8)

4. Mr. Hernandez bought $12\frac{3}{4}$ gallons of paint to paint his house. He used $10\frac{1}{4}$ gallons. How much paint was left?

   _____

5. The length of Dawn's yard is $8\frac{4}{5}$ feet. Find the width of her yard if it is $1\frac{3}{5}$ feet shorter than the length.

   _____

6. Find *eight and nine tenths minus three and four tenths.* Write your answer in words.

   _____

   _____

Name _____ Date _____

Homework Practice

Subtraction with Renaming

Subtract. Write each difference in simplest form.

1. $7\frac{1}{4} - 4\frac{3}{4} =$ _____

2. $9\frac{2}{5} - 5\frac{3}{5} =$ _____

3. $6\frac{1}{3} - 2\frac{2}{3} =$ _____

4. $14\frac{1}{2} - 5\frac{1}{4} =$ _____

5. $10\frac{5}{8} - 6\frac{6}{8} =$ _____

6. $12\frac{1}{5} - 6\frac{8}{10} =$ _____

7. $5\frac{1}{2} - 4\frac{5}{6} =$ _____

8. $3\frac{1}{3} - 1\frac{2}{3} =$ _____

9. $8\frac{4}{7} - 2\frac{6}{7} =$ _____

10. $3\frac{1}{4} - 1\frac{5}{8} =$ _____

11. $9\frac{8}{12} - 3\frac{11}{12} =$ _____

12. $2\frac{1}{10} - 1\frac{2}{5} =$ _____

13. $15\frac{5}{9} - 8\frac{7}{9} =$ _____

14. $6\frac{7}{16} - 2\frac{6}{8} =$ _____

Spiral Review

Use any strategy to solve each problem. (Lesson 10–9)

- Make a graph.
- Determine reasonable answers.
- Act it out.
- Look for a pattern.

15. At a grocery store a bag of apples costs $1.79. A jar of jelly costs $0.25 less than a bag of apples. Find the total cost of these two items.

_____

16. A runner starts running 10 miles per week and adds $\frac{1}{2}$ mile each week. How far will she run in the seventh week?

_____

11–1

Homework Practice

Units of Length

Estimate and then measure the length of each object. Find the measurement to the nearest $\frac{1}{4}$ inch or $\frac{1}{8}$ inch as shown.

1. to the nearest $\frac{1}{4}$ in.

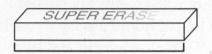

Estimate: _____

Measurement: _____

2. to the nearest $\frac{1}{8}$ in.

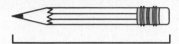

Estimate: _____

Measurement: _____

Choose an appropriate unit for measuring each length. Write *inch*, *foot*, *yard*, or *mile*.

3. length of a classroom _____

4. length of a pencil _____

5. distance between two cities _____

6. length of a football field _____

7. thickness of a book _____

8. width of Atlantic Ocean _____

Complete.

9. 4 ft = _____ in.

10. $1\frac{1}{2}$ yd = _____ in.

11. 15 ft = _____ yd

12. 3 yd = _____ in.

13. 5 mi = _____ ft

14. 40 in. = _____ ft

15. 180 in. = _____ yd

16. $2\frac{1}{3}$ yd = _____ ft

17. 4 mi = _____ yd

18. $\frac{1}{2}$ mi = _____ ft

19. $5\frac{1}{2}$ ft = _____ in.

20. 5 ft 11 in. = _____ in.

Spiral Review

Subtract. Write each difference in simplest form. Check your answer by drawing a picture

21. $8\frac{3}{10} - 5\frac{7}{10}$ _____

22. $6\frac{1}{4} - 2\frac{3}{4}$ _____

23. $14 - 3\frac{1}{3}$ _____

24. $12 - 9\frac{1}{4}$ _____

Name _____ Date _____

Homework Practice

Problem-Solving Strategy: Draw a Diagram

Solve. Use the *draw a diagram* strategy.

1. At a party, everyone shook hands with everyone else exactly once. There were a total of 36 handshakes. How many people were at the party?

   _____

2. Joey ate $2\frac{1}{4}$ eggs for breakfast on Monday. If he has eaten $13\frac{1}{2}$ eggs by the end of the week, how many eggs did he eat from Tuesday through Sunday? Write an equation and solve.

   _____

3. On the way to school, Zoe walks 3 blocks east, 2 blocks south, and 1 block west. What direction is the school from Zoe's home?

   _____

4. The Robinson family is having their portrait taken sitting on a bench. If Mr. and Mrs. Robinson sit on either end, how many different ways can their 3 children sit between them?

   _____

 **Review**

Choose an appropriate unit to measure the length of each.
Write *inch, foot, yard,* or *mile*. (Lesson 11–2)

5. distance between two countries

   _____

6. length of a football field

   _____

7. length of a necklace

   _____

8. width of a dresser

   _____

9. height of a 5th grader

   _____

10. width of computer

   _____

Name _____ Date _____

Homework Practice

Units of Weight

Complete.

1. 16 lb = _____ oz

2. 800 oz = _____ lb

3. 4 T = _____ lb

4. 126 lb = _____ oz

5. 16,000 lb = _____ T

6. 2,000 oz = _____ lb

7. 43 lb = _____ oz

8. 2,000 lb = _____ T

9. 3 lb = _____ oz

10. 144 oz = _____ lb

11. 50 lb = _____ oz

12. 18,000 lb = _____ T

Replace ( ) with <, >, or = to make a true statement.

13. 34 lb ( ) 534 oz

14. 290 lb ( ) 4,640 oz

15. 5 T ( ) 9,000 lb

16. 160 oz ( ) 10 lb

17. 400 oz ( ) 1 T

18. 150 lb ( ) 240 oz

19. 28 lb ( ) 450 oz

20. 19 lb ( ) 300 oz

Spiral Review

Solve. Use the *draw a diagram* strategy. (Lesson 11–2)

21. At a business meeting, everyone shook hands with everyone else exactly once. If there were a total of 10 handshakes, how many people were at the meeting?

_____

22. Amelia has 4 different pictures of her friends that she wants to hang on her wall. In how many different ways can she arrange the pictures?

_____

11-4

Homework Practice

Units of Capacity

Complete.

1. 6 c = _____ fl oz

2. 48 qt = _____ gal

3. 60 pt = _____ qt

4. 96 fl oz = _____ pt

5. 16 qt = _____ gal

6. 32 fl oz = _____ pt

7. 72 qt = _____ gal

8. 5 c = _____ fl oz

9. 22 c = _____ fl oz

10. 64 fl oz = _____ c

11. 52 pt = _____ c

12. 44 qt = _____ gal

Replace $\bigcirc$ with <, >, or = to make a true statement.

13. 64 fl oz $\bigcirc$ 7 c

14. 4 gal $\bigcirc$ 8 pt

15. 2 qt $\bigcirc$ 1 gal

16. 32 fl oz $\bigcirc$ 4 c

17. 9 c $\bigcirc$ 70 fl oz

18. 5 pt $\bigcirc$ 80 fl oz

19. 18 c $\bigcirc$ 6 pt

20. 12 qt $\bigcirc$ 4 gal

Spiral Review

Complete. (Lesson 11–3)

21. 128 oz = _____ lb

22. 16 lb = _____ oz

23. 12,000 lb = _____ T

24. 3 T = _____ lb

25. 48 oz = _____ lb

26. 240 oz = _____ lb

27. 59 lb = _____ oz

28. 4,000 lb = _____ T

29. 8 T = _____ lb

30. 80 oz = _____ lb

Name _____ Date _____

Homework Practice

Units of Time

Complete.

1. 4 wk = _____ days

2. 180 s = _____ min

3. 10 y = _____ mo

4. 3 d = _____ h

5. 4 min = _____ s

6. 5 h = _____ min

7. 10 min = _____ s

8. 36 wk = _____ d

9. 120 s = _____ min

10. 50 mo = _____ y _____ mo

11. 2 h = _____ s

12. 2 y = _____ mo

13. 30 min = _____ s

14. 3 d = _____ min

15. 4 wk = _____ h

16. 250 s = _____ min _____ s

17. 78 h = _____ d _____ h

18. 375 min = _____ h _____ min

Spiral Review

Complete. (Lesson 11–4)

19. 11 c = ▢ _____ fl oz

20. 16 qt = ▢ _____ gal

21. 50 c = ▢ _____ pt

22. 3 gal = ▢ _____ qt

23. 8 pt = ▢ _____ qt

24. 24 c = ▢ _____ pt

11-6

Homework Practice

Problem-Solving Investigation: Choose the Best Strategy

Use any strategy to solve each problem.

1. A number is divided by 10. Next the quotient is multiplied by 5. Then 15 is added to the product. If the result is 65, what is the number?

_____

2. Marta is 6 inches taller than Jennifer. Jennifer is 4 inches taller than Tara. Tara is 8 inches shorter than Jacqueline. If Jacqueline is 48 inches tall, how tall is Marta?

_____

3. Patrick is saving money for a new pair of shoes that costs $87. He has been saving his weekly allowance of $5 for 9 weeks. He also saved $22 from his birthday money. How much more money does Patrick need to save?

_____

4. Anna and Robert are putting together balloon bouquets with red and blue balloons. For every each bouquet includes 6 red balloons and $\frac{1}{3}$ as many blue balloons. If they have used 10 blue balloons, how many bouquets have they made in all?

_____

Spiral Review

Complete. (Lesson 11–5)

5. 480 s = ⬜ min _____

6. 15 min = ⬜ s _____

7. 288 h = ⬜ d _____

8. 30 h = ⬜ min _____

9. 4 mo = ⬜ w _____

10. 120 s = ⬜ min _____

Name _____ Date _____

Homework Practice

Elapsed Time

Find each elapsed time.

1. 10:15 P.M. to 10:59 P.M.　　**2.** 1:40 P.M. to 8:55 P.M.　　**3.** 9:25 A.M. to 8:20 P.M.

_____　　_____　　_____

4. 3:45 P.M. to 1:30 A.M.　　**5.** 2:26 A.M. to 8:00 A.M.　　**6.** 4:11 P.M. to 6:15 P.M.

_____　　_____　　_____

7. 12:09 P.M. to 2:00 P.M.　　**8.** 7:00 P.M. to 10:13 P.M.　　**9.** 5:55 A.M. to 6:30 P.M.

_____　　_____　　_____

10. 1:45 P.M. to 1:45 A.M.　　**11.** 4:22 A.M. to 7:40 A.M.　　**12.** 3:30 P.M. to 9:21 P.M.

_____　　_____　　_____

13. 7:12 P.M. to 8:55 P.M.　　**14.** 2:15 P.M. to 8:36 P.M.　　**15.** 1:11 P.M. to 3:47 P.M.

_____　　_____　　_____

16. 6:15 A.M. to 8:20 P.M.　　**17.** 4:48 P.M. to 12:01 A.M.　　**18.** 11:34 A.M. to 11:59 A.M.

_____　　_____　　_____

Spiral Review

Use any strategy to solve each problem. (Lesson 11–6)

19. Aaron is saving money to purchase a new pair of rollerblades. So far he has saved $9 a week for the past 5 weeks. If the rollerblades cost $55, how much more does Aaron need to save?

_____

20. A vase holds 70 fluid ounces of water minus 6 fluid ounces. How many quarts of water does the vase hold?

_____

21. Riley has 2 gallons of punch. She is filling glasses that hold 8 fluid ounces. How many glasses can be filled with 2 gallons of juice?

_____

22. Janna's class took a trip to the water park. They traveled 120 miles and stopped for a snack. They drove 120 more miles to the water park. Later in the day they traveled 64 miles and stopped for dinner. They drove 176 more miles home. Estimate the total number of miles Janna's class traveled.

_____

Homework Practice

Units of Length

Complete.

1. 26 cm = _____ mm

2. 700 cm = _____ m

3. 8 km = _____ m

4. 0.6 m = _____ cm

5. 4,000 mm = _____ m

6. 250 mm = _____ cm

7. 800 cm = _____ mm

8. 23 cm = _____ mm

9. 0.25 km = _____ m

10. 300 cm = _____ m

11. 6 m = _____ cm

12. 3,000 m = _____ km

13. 6 cm = _____ mm

14. 5 km = _____ m

Solve.

15. Which is a more reasonable estimate for the depth of a swimming pool: 10 millimeters, 10 meters, or 10 kilometers?

16. When completed, a tunnel will be 1.3 km long. What is this length in meters?

Spiral Review

Find each elapsed time. (Lesson 11–7)

17. 6:29 A.M. to 7:46 A.M. _____

18. 11:09 A.M. to 12:05 P.M. _____

19. 4:16 P.M. to 6:21 P.M. _____

20. 5:30 P.M. to 7:19 P.M. _____

Name _____ Date _____

Homework Practice

Problem-Solving Strategy: Determine Reasonable Answers

Is each estimate reasonable? Explain.

1. Jamil volunteers once a week. He works for 3.75 hours at a time. Is 40 hours a reasonable estimate for how many hours he will work in 10 weeks?

_____

2. Gamal collects cards. He buys about 4 cards a week. Is 800 cards a reasonable estimate for how many cards he will buy in one year?

_____

3. Kim invited 5 friends over to swim. They took turns on the 3 rafts. If they each lay on a raft for 30 minutes at a time, is 2 hours a reasonable estimate for how long would it take for all 5 friends to have their turn?

_____

4. The Ling family ordered 3 hamburgers, 2 fries, and 3 drinks. If they paid with three 10-dollar bills, is $3.00 a reasonable estimate for the amount of change they will receive?

Item	Cost
Hamburger	$3.95
Fries	$2.90
Drink	$2.85

_____

Spiral Review

Complete.   (Lesson 12–1)

5. 4 km = ▇ m _____

6. 600 cm = ▇ m _____

7. 10 mm = ▇ cm _____

8. 5 m = ▇ cm _____

Name _____ Date _____

Homework Practice

Units of Mass

Complete.

1. 90 g = _____ kg

2. 300 g = _____ kg

3. 1,000 mg = _____ g

4. 0.9 kg = _____ g

5. 5 g = _____ kg

6. 0.004 kg = _____ g

7. 25 kg = _____ g

8. 670 g = _____ kg

Replace ◯ **with <, >, or = to make a true statement.**

9. 2.4 g ◯ 240 mg

10. 8 kg ◯ 80,000 g

11. 1.32 g ◯ 1,320 mg

12. 510 mg ◯ 5.1 g

13. 3,500 mg ◯ 35 g

14. 370 mg ◯ 3.7 g

Solve.

15. A box of pasta has a mass of 454 grams. How many boxes should Leo buy if he wants to cook at least 1 kilogram of pasta? Explain.

_____

Spiral Review

Is the estimate reasonable? Explain.   (Lesson 12–2)

16. Miriam's computer weighs 165 ounces. She estimates that it weighs about 20 pounds. Is Miriam's estimate reasonable?

_____

17. Avner needs to buy 12 yards of fabric. At the store, all of the fabric is marked in feet. Avner estimates that 40 feet of fabric will be long enough. Is his estimate reasonable?

_____

Name _____ Date _____

Homework Practice

Units of Capacity

Complete.

1. 7,200 mL = _____ L

2. 490 mL = _____ L

3. 0.1 L = _____ mL

4. 7,000 mL = _____ L

5. 3 L = _____ mL

6. 8 mL = _____ L

7. 9,000 mL = _____ L

8. 0.53 L = _____ mL

Replace each ⬭ with <, >, or = to make a true statement.

9. 6.4 L ⬭ 640 mL

10. 5 L ⬭ 50,000 mL

11. 2.32 L ⬭ 2,320 mL

12. 410 mL ⬭ 4.1 L

13. 1,500 mL ⬭ 15 L

14. 970 mL ⬭ 9.7 L

Solve.

15. Tracy has a 5-liter punch bowl. She buys two containers of juice that hold 1.75 liters and 2.7 liters. Can she empty the two containers into the bowl? Explain.

_____

_____

Spiral Review

Complete. (Lesson 12–3)

16. 1 g = ▨ mg _____

17. 350 g = ▨ kg _____

18. 4,600 g = ▨ kg _____

19. 1 kg = ▨ g _____

Name _____ Date _____

Homework Practice

Integers and Graphing on Number Lines

Write an integer to represent each situation. Then graph the integer on a number line.

1. a loss of $10 _____

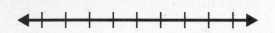

2. 18 feet above sea level _____

3. add 2 quarts of punch _____

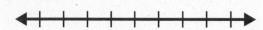

4. 6 marbles are taken away _____

Write an integer to represent each situation. Then write its opposite.

5. The temperature rose 5 degrees.

_____

6. Greg lost $4 on the way to school.

_____

7. Justine grew 2 inches last year.

_____

8. Robby withdrew $10 from the bank.

_____

Spiral Review

Complete.   (Lesson 12–4)

9. 80 L = ▮ mL _____

10. 6 L = ▮ mL _____

11. 426 mL = ▮ L _____

12. 0.06 L = ▮ mL _____

Name _____ Date _____

Homework Practice

Units of Temperature

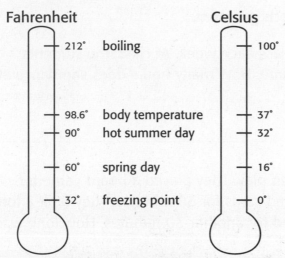

Choose the more reasonable temperature for each situation. Use the thermometers above if needed.

1. bath water: 75°F or 105°F _____

2. frozen yogurt: −10°C or 10°C _____

3. ice skating rink: 25°F or 50°F _____

4. pie in an oven: 100°F or 300°F _____

Find each change in temperature. Use an integer to represent the change.

5. 43°C to 22°C _____

6. 61°F to 79°F _____

7. 54°F to 32°F _____

8. 24°C to 51°C _____

Spiral Review

Write an integer to represent each situation.   (Lesson 12–5)

9. found $2 _____

10. 36 degrees below 0 _____

11. 6 points added to a grade _____

12. 3 yards lost in football _____

13. lost $3 _____

14. earned $20 working _____

Name _____ Date _____

Homework Practice

Problem-Solving Investigation: Choose the Best Strategy

Use any strategy to solve each problem.

1. Hoshi attends her ballet class each week. At class, the students dance for 2.3 hours at a time. How many hours does she dance at class in 20 weeks?

   _____

2. Seki had her friends over to play. They played a board game for 45 minutes and then played cards for 30 minutes. They built a fort for 45 minutes and painted for another 30 minutes. How long was the play date?

   _____

3. Jack ordered 3 drums, 2 blankets, and 3 pants. If he paid with eight 20-dollar bills, how much change will he get back?

   _____

Item	Cost
Blanket	$15.95
Pants	$12.99
Drum	$24.95

 **Spiral Review**

Find each change in temperature. Use an integer to represent the change.

4. −5°C to 0°C _____

5. 38°F to 56°F _____

6. 95°F to 87°F _____

7. 13°C to 8°C _____

8. −4°C to 12°C _____

9. 14°F to −1°F _____

Name _____ Date _____

Homework Practice

Geometry Vocabulary

Use the figure to determine if each pair of lines is *parallel, intersecting,* or *perpendicular.* Choose the most specific term.

1. $\overline{AB}$ and $\overline{CD}$

_____

2. $\overline{BD}$ and $\overline{CD}$

_____

3. $\overline{AD}$ and $\overline{CD}$

_____

Describe each figure below as a *point, line, ray* or *line segment.*

4.

_____

5.

_____

6.

_____

7.

_____

Spiral Review

Use any strategy to solve each problem.

8. Judy ran a 3-kilometer race. When she was halfway to the finish line, how far had she run?

_____

9. A museum charges $3 admission. If the museum collected $1,176, how many people came to the museum?

_____

Homework Practice

Problem-Solving Strategy: Use Logical Reasoning

1. Mika and Pazi each think of a number. Mika's number is 7 more than Pazi's number. The sum of the two numbers is 49. What is Pazi's number?

_____

2. A park has an area of 64 square meters and a perimeter of 40 meters. Find the length and width of the park.

_____

3. Breanna has quarters, dimes, and nickels in her purse. She has 3 fewer nickels than dimes, but she has 2 more nickels than quarters. If Breanna has 2 quarters, how much money does she have?

_____

4. Jennifer, Tara, and Brooke are waiting in a line. Brooke is not first in line. Jennifer is behind the oldest in line. Brooke is behind Jennifer. List the girls in order from first to last.

_____

Measure each segment. Then determine whether each pair of line segments are congruent. Write *yes* or *no*.

5.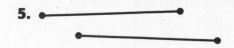

6.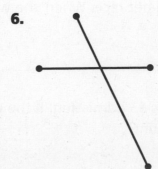

_____   _____

Name _____ Date _____

Homework Practice

Triangles

Classify each triangle drawn or having the given angle measures as *acute*, *right*, or *obtuse*.

1.

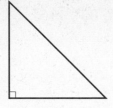

_____

2.

_____

3. 70°, 60°, 50°

_____

The sum of the measures of the angles of a triangle is 180°. Find the value of x in each triangle. Then classify each triangle as *scalene*, *isosceles*, or *equilateral*.

4.

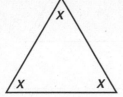

_____

_____

5.

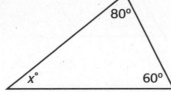

_____

_____

6.

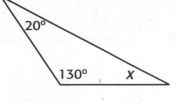

_____

_____

Spiral Review

Solve. Use the logical reasoning strategy. (Lesson 13–2)

7. In August Daryl ran 3 miles every other day. In September, he ran 3.5 miles every other day. If the trend continues, how much will he run each day in October?

_____

Name _____ Date _____

Homework Practice

Quadrilaterals

Find the value of x in each quadrilateral. The sum of the measures of the angles of a quadrilateral is 360°.

1.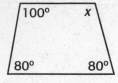

2.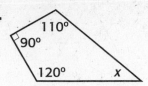

3. 75°, 85°, 115°, x

_____   _____   _____

Classify each quadrilateral using the best description.

4.

5.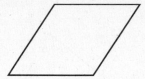

6.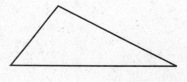

_____   _____   _____

Spiral Review

Find the number of congruent sides in each triangle. Then state whether any of the sides appear to be perpendicular. Write *yes* or *no*. (Lesson 13–3).

7.

8.

_____   _____

Name _____ Date _____

Homework Practice

Problem-Solving Investigation: Choose the Best Strategy

Use any strategy shown below to solve each problem.

- Look for a pattern
- Draw a diagram
- Guess and check

Use the picture to answer Exercises 1–3.

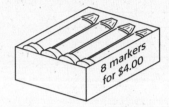 

1. Compare the 2 containers of markers. Which is the better buy?

_____

2. If you bought 1 of the first box and 3 of the second box, and you gave the cashier three $5 bills, how much change would you get back?

_____

3. You buy four boxes of markers and it costs you $16. Which kind did you buy?

_____

Name the quadrilaterals that have the given characteristics. (Lesson 13–4)

4. exactly one pair of parallel sides

5. All adjacent sides are perpendicular

_____   _____

Name _____ Date _____

Homework Practice

Translations and Graphs

Graph each figure and the translation described. Write the ordered pairs for the new vertices.

1. triangle *FGH* with vertices *F*(2, 1), *G*(5, 4), *H*(6, 0); translated 2 units right, 4 units up

   _____

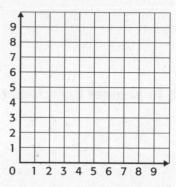

2. quadrilateral *JKLM* with vertices *J*(4, 1), *K*(5, 5), *L*(8, 5), *M*(7, 1); translated 1 unit left, 3 units up

   _____

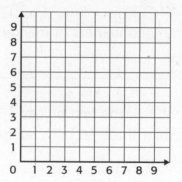

Spiral Review

Use any strategy shown below to solve. (Lesson 13–5)

   • Write an equation   • Look for a pattern   • Use logical reasoning

3. There are 5 pecans for every 7 cashews in a bowl of mixed nuts. If there are 45 pecans in the bowl, how many cashews are there?

   _____

4. Tess has more books than Brett but fewer than Halley. If Halley has fewer books than Molly, who has the fewest books?

   _____

Name _____ Date _____

Homework Practice

Reflections and Graphs

Graph each figure after a reflection across the line. Then write the ordered pairs for the new vertices.

1.

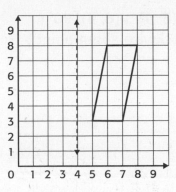

_____

2.

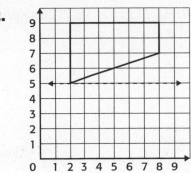

_____

Spiral Review

Graph each figure after the translation described. Write the ordered pairs for the new vertices.

3. quadrilateral *ABCD* with vertices *A*(4, 7), *B*(7, 7), *C*(9, 2), *D*(6, 2); translated 3 units left

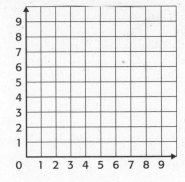

_____

4. triangle *FGH* with vertices *F*(1, 5), *G*(4, 7), *H*(4, 2); translated 4 units right, 1 unit up

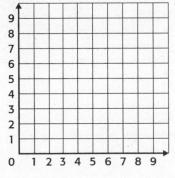

_____

103

Name _____ Date _____

Homework Practice

Rotations and Graphs

Graph each triangle with the given vertices and the rotation described. Write the ordered pairs for the new vertices.

1. *J*(1, 6), *K*(4, 6), *L*(2, 2); 180° clockwise about point *K*

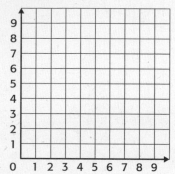

2. *T*(8, 8), *U*(8, 4), *V*(6, 4); 90° counterclockwise about point *V*

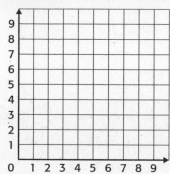

_____   _____

Spiral Review

Graph each figure after a reflection across the line. Then write the ordered pairs for the new vertices. (Lesson 13–7)

3.

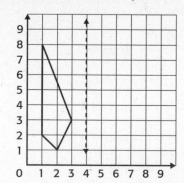

4.

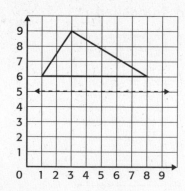

_____   _____

Name _____ Date _____

Homework Practice

Identify Transformations

Determine whether each transformation is a <u>translation</u>, <u>reflection</u>, or <u>rotation</u>.

1.

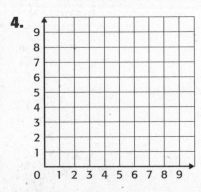

_____

2.

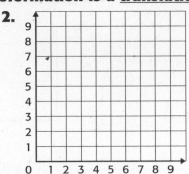

_____

3.

_____

4.

_____

5.

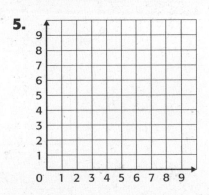

_____

6.

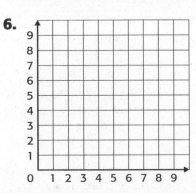

_____

Spiral Review

Graph each triangle with the given vertices and the rotation described. Write the ordered pairs for the new vertices. (Lesson 13-8)

7. Q(2, 1), R(2, 5), S(5, 4); 180°
clockwise about point S

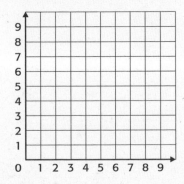

_____

8. T(1, 7), U(2, 1), V(4, 7); 90°
counterclockwise about point V

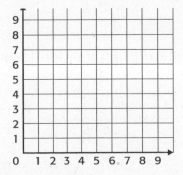

_____

Name _____ Date _____

Homework Practice

Perimeters of Polygons

Find the perimeter of each square or rectangle.

1.   13 ft

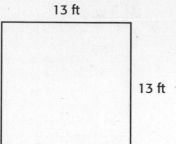

13 ft

_____

2.   4.76 m

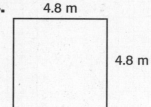

1.93 m

_____

3.   11 ft

$2\frac{1}{2}$ ft

_____

4.   4.8 m

4.8 m

_____

5. Neil made a wooden, rectangular picture frame that is 14 inches long and 10 inches wide. If he charges $2.50 per foot, how much will he sell this frame for?

_____

Spiral Review

(Lesson 13–9)

6. Create a pattern using transformations.

_____

Name _____ Date _____

Homework Practice

Area

Estimate the area of each figure. Each square represents 1 square centimeter.

1.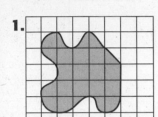

2.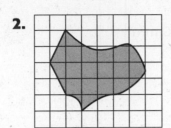

3.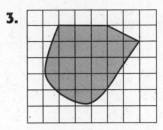

_____

_____

_____

4.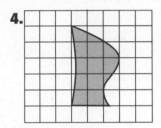

5.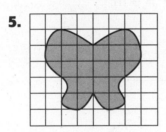

6.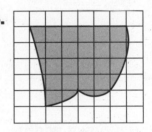

_____

_____

_____

Spiral Review

Find the perimeter of each square or rectangle.   (Lesson 14–1)

7. 3 m / 3 m

8. 4 in. / 9 in.

9. 3 yd / 7 yd

_____

_____

_____

Name _____ Date _____

Homework Practice

Areas of Rectangles and Squares

Find the area of each rectangle.

1.

2 cm
4 cm

_____

2.

40 mm
15 mm

_____

3.

4 in.
4 in.

_____

4. rectangle

$\ell = 3$ yd

$w = 4$ yd

_____

5. rectangle

$\ell = 4$ in.

$W = 5$ in.

_____

6. rectangle

$\ell = 32$ mm

$w = 46$ mm

_____

Find the unknown width.

7. rectangle

$\ell = 3$ in.

$A = 6$ square inches

$w =$ _____

8. rectangle

$\ell = 45$ mm

$A = 3{,}150$ square millimeters

$w =$ _____

Spiral Review

Solve.

9. Mike's room is 12 feet by 15 feet. How many square feet of carpeting does he need to cover the entire floor?

_____

10. Helen is planting tomatoes in her garden. She can place 3 plants per square foot. How many plants does she need if her garden measures 7 ft by 6 ft?

_____

Name _____ Date _____

Homework Practice

Three-Dimensional Figures

Describe parts of each figure that are perpendicular and congruent. Then identify the figure.

1.

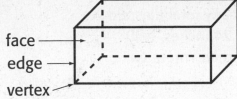

face →
edge →
vertex →

_____

_____

2.

_____

_____

3.

_____

_____

Spiral Review

Find the area of each rectangle. (Lesson 14–3)

4.

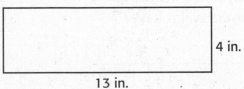

4 in.

13 in.

_____

5.

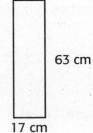

63 cm

17 cm

_____

6.

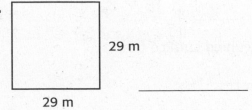

29 m

29 m

_____

Homework Practice

Problem-Solving Strategy: Make a Madel

Solve. Use the *make a model* strategy.

1. Nan and Sato are designing a coffee table using 4 inch tiles. Nan uses 30 tiles and Sato uses half as many. How many total tiles did they use? If the area of the table is 36 inches by 24 inches, will they have enough tiles for the table? If not, how many more will they need?

   _____

2. The Jones family is landscaping their yard. If they have a yard that is 160 square feet, and one side is 10 feet long, what is the length of the other side of the garden? If they plant 3 bushes that need to be 3 feet apart and 3 feet away from the fence around the yard, will they have the space?

   _____

3. Bob is organizing his closet. If he has clothing bins that measure 20 inches high, 18 inches wide, and 14 inches long, how many bins can he fit in a 60-inch long closet that is 30 inches deep and 72 inches high?

   _____

4. Roberto wants to build a brick wall. Each brick layer is 3 inches thick, and the wall will be 18 inches tall. How many layers will it have?

   _____

Spiral Review

Identify each figure.   (Lesson 14–4)

5. This polyhedron has six rectangular faces. _____

6. This prism has triangular bases. _____

7. This is a solid that has a circular base and one curved surface from the base to a vertex. _____

Name _____ Date _____

Homework Practice

Volume of Prisms

Find the volume of each prism.

1.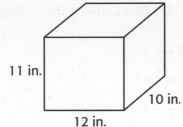

11 in.
10 in.
12 in.

_____

2.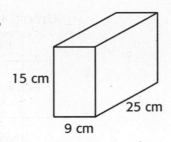

15 cm
25 cm
9 cm

_____

3.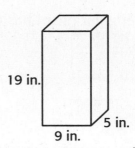

19 in.
5 in.
9 in.

_____

4. What is the volume of a rectangular box that has a base of 50 square inches and a height of 12 inches?

_____

5. Bernice made a rectangular wooden tool box that has a base of 50 square centimeters and a height of 35 cm. What is the volume?

_____

Spiral Review

Use any strategy to solve. (Lesson 15–5)

6. Ali has a loaf of bread that he needs to slice for his family's dinner. How many cuts does he need to make if he needs 6 equal-size slices of bread?

_____

7. Maggie's older sister is repaying her student loans. Her loans, including interest, total $9,985. How much are her monthly payments if she plans to repay the loans in 8 years?

_____

Name _____ Date _____

Homework Practice

Surface Areas of Prisms

The **surface srea** (*SA*) of a 3-dimensional figure is the sum of the area of all its faces.

A rectangular prism has 6 faces.

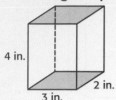

Unfold the prism to examine the 6 faces.

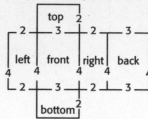

Find the area of each of the 6 faces, and add.

Face	Area	In.²
front	3 × 4	12
back	3 × 4	12
top	3 × 2	6
bottom	3 × 2	6
left	2 × 4	8
right	2 × 4	8
	Total	52

The surface area of this rectangular prism is 52 in.²

Find the surface area of each figure.

1.

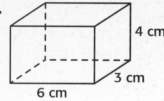

_____

2.

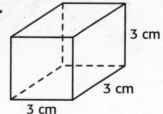

_____

3.

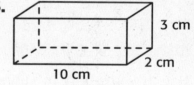

_____

Spiral Review

Find the volume of each prism. **(Lesson 14–6)**

4.

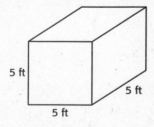

5.

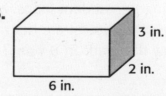

6.

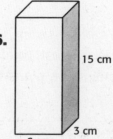

_____   _____   _____

Name _____ Date _____

Homework Practice

Select Appropriate Measurement Formulas

Determine whether you need to find the perimeter, area, or volume. Then solve.

1. Charlene is filling a box planter with soil. Each bag of soil contains enough to fill 500 cubic inches. If the base of the box planter is a 9-inch square and the sides are 18 inches tall, how many bags of soil will Charlene need?

_____

2. Tobias is helping his uncle paint the side of a barn. Each can of paint covers 80 square feet. If the side of the barn is 10 feet high and 15 feet wide, how many cans of paint will Tobias need?

_____

3. Gina is tying up a stack of newspapers with string. She wants the string to wrap around twice, once lengthwise and once crosswise. If the stack of newspapers is 11 inches wide, 13 inches long, and 15 inches high, how much string will she need?

_____

Spiral Review

Find the surface area of each prism. (Lesson 14–7)

4.
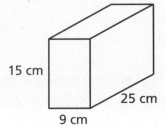
15 cm
25 cm
9 cm

5.
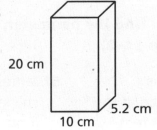
20 cm
5.2 cm
10 cm

6.
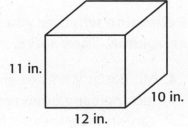
11 in.
10 in.
12 in.

_____   _____   _____

Name _____ Date _____

Homework Practice

Problem-Solving Investigation: Choose the Best Strategy

Use any strategy shown below to solve each problem.

- Make a model
- Draw a diagram
- Look for a pattern
- Use logical reasoning

1. The Humane Society is building new cages for their dogs and cats. They have 2 crews of workers building them. There are 28 dogs and 34 cats. All the animals are kept separate. The first crew can build a cage in 1 hour and the second crew, which is smaller, takes 2 hours to build a cage. How many hours will it take to build the cages using both crews if they do not take a break?

_____

2. You are shopping for some new clothes. You buy a shirt for $28 and a pair of dress shoes for $45. If you give the cashier a $100 bill, how much change will you get back?

_____

3. Marge needs to bake 8 dozen cookies for a bake sale. For each batch of cookies she needs $4\frac{1}{2}$ cups of flour. Each batch makes 2 dozen cookies. How much flour does she need?

_____

Spiral Review

Determine whether you need to find the perimeter, area, or volume. Then solve. (Lesson 14–8)

4. Mr. Bauer wants to enclose his rectangular garden with a fence. If the garden measures 12 feet by 9 feet, how many feet of fencing will he need to buy?

_____

5. Jameson has a storage bin that is the shape of a cube for his building blocks. If one side of the cube measures 2 feet, how many cubic feet of space does he have for storage?

_____

Name _____ Date _____

Homework Practice

Probability

Suppose you spin the spinner at right. Describe the probability of landing on each pattern. Write *certain, impossible, unlikely, equally likely,* or *likely*.

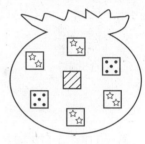

1. striped, plain, or speckled _____

2. striped _____

3. speckled _____

One cube is drawn without looking. Describe the probability of choosing a pattern. Write *certain, impossible, unlikely, equally likely,* or *likely*.

4. five-star _____

5. spots _____

6. two star _____

Spiral Review

Determine whether you need to find the perimeter, area, or volume. Then solve.   (Lesson 14–7)

7. Elizabeth is planting a garden. She wants to use a space that is 5 yards long and 3 yards wide. How much space will she have in her garden?

_____

8. Which units would be most appropriate to measure the volume of a pool: cubic inches, cubic feet, or cubic yards? Explain.

_____

Name _____ Date _____

Homework Practice

Probability as a Fraction

The bag shown contains cubes with dots, stars, and stripes. Find the probability of each event. Write the answer as a fraction in simplest form.

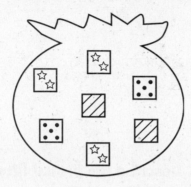

1. P(star)

_____

2. P(stripe)

_____

3. P(not a stripe)

_____

4. P(star or stripe)

_____

List the possible outcomes in each probability experiment. **(Lesson 15–1)**

5. rolling a number cube

_____

6. randomly choosing a letter from the word *PROBABILITY*.

_____

Name _____ Date _____

Homework Practice

Problem-Solving Strategy: Make an Organized List

Solve by *making an organized list*.

1. Andy only knows three people in the study hall. Desks are arranged in pairs. How many possible ways can Andy sit next to someone he knows?

   _____

2. Russ has to go to the office, the school store, and the water fountain. How many different ways can Russ make the stops?

   _____

3. Linda has black pants and a pair of jeans, black and red shoes, a red striped jersey and a white jersey. How many outfits can she make if she always wears a jersey, pants, and shoes?

   _____

4. How many different ways can you write the product of the prime factors of 24?

   _____

Spiral Review

A number cube with sides labeled 1 through 6 is rolled. Find the probability of each event. Write as a fraction in simplest form.   (Lesson 15–2)

5. $P(4)$ _____

6. $P(\text{a number less than 4})$ _____

7. $P(\text{even number})$ _____

8. $P(7)$ _____

Name _____ Date _____

15-4

Homework Practice

Counting Outcomes

Use a tree diagram to find the possible outcomes.

1. How many choices do you have for your lunch if you pick either ham or roast beef with cheese, tomatoes, or onions?

   _____

2. You go to a playground. You decide to climb across the monkey bars, go down the slide, and climb the rock wall. How many different ways can you complete the activities?

   _____

3. You are getting ready for school, and you only have a choice of white, black, or yellow shoes and either a pair of jeans or shorts. How many possible combinations can you have?

   _____

Use the spinner below to answer Exercises 4 and 5. Find the probability of each event. Write the answer as a fraction in simplest form. (Lesson 15–2)

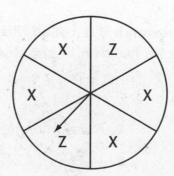

4. P(X)   5. P(Z)

_____   _____

Name _____ Date _____

Homework Practice

Problem-Solving Investigation: Choose the Best Strategy

Use any strategy shown below to solve each problem.

• Look for a pattern.   • Work backward.   • Solve a simpler problem.

Use the following Venn diagram for Exercises 1–3.

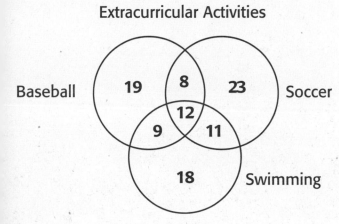

1. How many total people played soccer?

   _____

2. How many people played only soccer and baseball?

   _____

3. How many swam and played baseball only?

   _____

4. How many played baseball, swam, and played soccer?

_____

5. How many people swam and played soccer?

_____

Spiral Review

A coin is tossed twice. (Lesson 15–4)

6. Make a tree diagram to show all possible outcomes.

7. What is the probability of tossing heads, then tails? _____

8. What is the probability of tossing the same thing in a row? _____